THE CONTEST PROBLEM BOOK VI

American High School Mathematics Examinations 1989–1994

First Printing

Published in Washington, DC by
The Mathematical Association of America

Library of Congress Catalog Card Number 00-101675

Complete Set ISBN: 0-88385-600-X

Vol. 31 ISBN: 0-88385-642-5

Manufactured in the United States of America

THE CONTEST PROBLEM BOOK VI

American High School Mathematics Examinations 1989–1994

Compiled and augmented by

Leo J. Schneider
John Carroll University

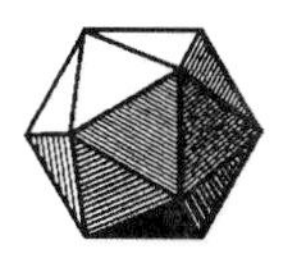

40

THE MATHEMATICAL ASSOCIATION OF AMERICA

ANNELI LAX NEW MATHEMATICAL LIBRARY

PUBLISHED BY

THE MATHEMATICAL ASSOCIATION OF AMERICA

The New Mathematical Library (NML) was started in 1961 by the School Mathematics Study Group to make available to high school students short expository books on various topics not usually covered in the high school syllabus. In a decade the NML matured into a steadily growing series of some twenty titles of interest not only to the originally intended audience, but to college students and teachers at all levels. Previously published by Random House and L. W. Singer, the NML became a publication series of the Mathematical Association of America (MAA) in 1975. Under the auspices of the MAA the NML continues to grow and remains dedicated to its original and expanded purposes. In its third decade, it contains forty titles.

ANNELI LAX NEW MATHEMATICAL LIBRARY

1. Numbers: Rational and Irrational *by Ivan Niven*
2. What is Calculus About? *by W. W. Sawyer*
3. An Introduction to Inequalities *by E. F. Beckenbach and R. Bellman*
4. Geometric Inequalities *by N. D. Kazarinoff*
5. The Contest Problem Book I Annual High School Mathematics Examinations 1950–1960. Compiled and with solutions *by Charles T. Salkind*
6. The Lore of Large Numbers *by P. J. Davis*
7. Uses of Infinity *by Leo Zippin*
8. Geometric Transformations I *by I. M. Yaglom, translated by A. Shields*
9. Continued Fractions *by Carl D. Olds*
10. Replaced by NML-34
11. } Hungarian Problem Books I and II, Based on the Eötvös Competitions
12. } 1894–1905 and 1906–1928, *translated by E. Rapaport*
13. Episodes from the Early History of Mathematics *by A. Aaboe*
14. Groups and Their Graphs *by E. Grossman and W. Magnus*
15. The Mathematics of Choice *by Ivan Niven*
16. From Pythagoras to Einstein *by K. O. Friedrichs*
17. The Contest Problem Book II Annual High School Mathematics Examinations 1961–1965. Compiled and with solutions *by Charles T. Salkind*
18. First Concepts of Topology *by W. G. Chinn and N. E. Steenrod*
19. Geometry Revisited *by H. S. M. Coxeter and S. L. Greitzer*
20. Invitation to Number Theory *by Oystein Ore*
21. Geometric Transformations II *by I. M. Yaglom, translated by A. Shields*
22. Elementary Cryptanalysis—A Mathematical Approach *by A. Sinkov*
23. Ingenuity in Mathematics *by Ross Honsberger*
24. Geometric Transformations III *by I. M. Yaglom, translated by A. Shenitzer*
25. The Contest Problem Book III Annual High School Mathematics Examinations 1966–1972. Compiled and with solutions *by C. T. Salkind and J. M. Earl*
26. Mathematical Methods in Science *by George Pólya*
27. International Mathematical Olympiads—1959–1977. Compiled and with solutions *by S. L. Greitzer*
28. The Mathematics of Games and Gambling *by Edward W. Packel*
29. The Contest Problem Book IV Annual High School Mathematics Examinations 1973–1982. Compiled and with solutions *by R. A. Artino, A. M. Gaglione, and N. Shell*
30. The Role of Mathematics in Science *by M. M. Schiffer and L. Bowden*
31. International Mathematical Olympiads 1978–1985 and forty supplementary problems. Compiled and with solutions *by Murray S. Klamkin*
32. Riddles of the Sphinx *by Martin Gardner*
33. U.S.A. Mathematical Olympiads 1972–1986. Compiled and with solutions *by Murray S. Klamkin*
34. Graphs and Their Uses *by Oystein Ore.* Revised and updated *by Robin J. Wilson*
35. Exploring Mathematics with Your Computer *by Arthur Engel*
36. Game Theory and Strategy *by Philip D. Straffin, Jr.*

37. Episodes in Nineteenth and Twentieth Century Euclidean Geometry *by Ross Honsberger*
38. The Contest Problem Book V American High School Mathematics Examinations and American Invitational Mathematics Examinations 1983–1988. Compiled and augmented *by George Berzsenyi and Stephen B. Maurer*
39. Over and Over Again *by Gengzhe Chang and Thomas W. Sederberg*
40. The Contest Problem Book VI American High School Mathematics Examinations 1989–1994. Compiled and augmented *by Leo J. Schneider*

Other titles in preparation.

Books may be ordered from:
MAA Service Center
P. O. Box 91112
Washington, DC 20090-1112
1-800-331-1622 fax: 301-206-9789

Contents

Preface

The annual American Mathematics Competitions are a program of the Mathematical Association of America for pre-college students. Great care is taken to create new and interesting problems for these competitions. After each examination, all the problems and their solutions are made public in individual pamphlets so students can learn mathematics by practicing for future contests, and to give teachers new and exciting problems for their classes.

The oldest of the American Mathematics Competitions is the American High School Mathematics Examination [AHSME], first given in 1949. This book presents the problems and solutions from the 40th through 45th annual AHSMEs, the examinations administered 1989 through 1994. All the statements of the problems appear in this book exactly as they appeared on the examinations.

The solutions in this book include all the official solutions originally made available after the competitions, many enhanced. In addition, this book includes many alternative solutions that have never been published before.

How To Use This Book

Students who will participate in the American Mathematics Competitions and many other mathematics contests can use the problems in this book for practice. The rules for the AHSME appear on the page xix, preceding the statements of the AHSME problems so students can try the examinations under simulated contest conditions. Answers follow the questions from each competition, together with the percentage of times honor roll students used each answer choice. A few comments calling attention to unusually attractive 'distractors' are also included. One or more complete solutions to each problem are given in the section following all the examination questions and answers.

Anyone interested in mathematics will find interesting and challenging problems in these examinations. Ignore the multiple choice nature of the AHSME questions, put your answer in simplest form, and compare your solution to the one in the solutions section.

Students can easily form their own lists of the mathematics most valuable in solving the types of problems that frequent mathematics competitions. The penultimate section in this book is a short discussion of problem-solving, and can be useful in starting this list.

A book like this is not made to be read cover to cover like a novel. Unless you are attempting the problems under simulated contest conditions, after trying some problems you will want to look at the answers. Be sure to study the solutions, even for those questions you answered correctly. Knowing several approaches to the same problem is valuable and useful.

Those who would like to look at how a particular topic may occur in some of the problems or solutions can examine the last section of this book. It attempts to classify problems and methods of solutions.

Special Notation

Standard mathematical notation is used for the questions and solutions in these examinations and in this book. To avoid ambiguity in the statements of the problems on the competitions, any notation the committee thinks might be unfamiliar to participants is explained in parenthetical comments. For example, this includes:

- $\lfloor v \rfloor$ which stands for the greatest integer less than or equal to the real number v.
- $\lceil v \rceil$ which stands for the least integer greater than or equal to the real number v.

For the geometry problems on our examinations and in this book, we distinguish notationally only between geometric objects and numbers:

- $\overline{AB}$ is used for either the *line* or the *ray* or the line *segment* through points A and B, the context clarifies which.
- AB stands for the *length* of the segment between A and B.
- $[P_1P_2 \dots P_n]$ stands for the *area* enclosed by the *polygon* $P_1P_2 \dots P_n$.

AHSME Honor Roll

The AHSME rules on page xix give the scoring rules. A student achieves the AHSME Honor Roll by scoring 100 or more points out of the 150 possible points. The Honor Roll students are those invited to take the American Invitational Mathematics Examination [AIME].

To acquire the 100 or more points for the Honor Roll requires the student to correctly answer about half of the 30 questions and skip the rest. The student who attempts more problems suffers severe penalties for wrong answers: To make up for every three wrong answers, two additional questions must be answered correctly. Where n and c represent the number of questions answered and the number of correct answers, respectively, it might be instructive to graph the lattice points in the n-c plane where

$$5c + 2(30 - n) \geq 100$$
$$30 \geq n \geq c \geq 0.$$

Note from the graph that $(14, 14)$ is the leftmost lattice point in the region. It represents an honor roll student who obtained a score of 102 by answering 14 questions and getting each one correct. The points on the line

$$5c + 2(30 - n) = 100,$$

$(n, c) = (15, 14)$, $(20, 16)$, $(25, 18)$, and $(30, 20)$, represent the students who scored exactly 100 points by answering 15 questions with 1 wrong, 20 questions with 4 wrong, 25 questions with 7 wrong, and 30 questions with 10 wrong, respectively.

The Difficulty of the AHSME

There is no pretesting of the questions to be used on any of the American Mathematics Competitions. The committees use their best judgement to construct the examination with some level of difficulty in mind.

For the six AHSMEs in this book, the goal was to have approximately 4000 honor roll students per year. A test discriminates most among those students who get approximately half right. It is for this reason, and not because the members of the committee are mathematically sadistic, that the goal was to have between one and two percent of the participants get the equivalent of more than half the problems correct.

Any experienced teacher knows that test questions thought to be easy can actually be quite challenging for the students, and vice versa. Each misjudgement of the difficulty of one of the 30 AHSME questions usually affects the number of honor roll students by a factor of 2. Therefore, the committee would have been quite happy had there been anywhere between 2000 and 8000 honor roll students for each test. In fact, the statistics detailed on the pages just following each set of AHSME questions shows that broader goal was missed more often than not, but most of the results of the examinations were somewhat close to it.

The committee was well-aware that the number of participants dropped by over 13% from the 40th AHSME to the 44th AHSME, and this drop was not totally accounted for by the demographics of a decreasing high school population. For most of these AHSMEs, the committee had missed its honor-roll goal on the low side. Therefore, a determined effort was made to design the 45th AHSME to yield more honor roll students. A glance at the statistics for that test shows it was more successful in that effort than the committee ever dreamed it would be!

Related Mathematics Competitions

Besides the AHSME discussed in this book, the American Mathematics Competitions include the American Junior High School Mathematics Examination [AJHSME], the American Invitational Mathematics Examination [AIME], and the United States of America Mathematical Olympiad [USAMO]. The AJHSME is a 25-question, 40-minute, multiple-choice examination for students in the eighth grade or lower. The AIME is a 15-question, 3-hour examination with each answer a three-digit non-negative integer. The USAMO is a five-question, 3.5-hour, free-response examination for about 150 students with the highest combined scores on the AHSME and AIME. The six students with the highest combined scores on the AHSME, AIME and USAMO form the team that represents the United States in the annual International Mathematical Olympiad.

Related Problem Solving References

Volumes in the New Mathematics Library published by the Mathematical Association of America that contain problems related to the American Mathematics Competitions are:

Vol. 5, The Contest Problem Book I, AHSME 1950–60.

Vol. 17, The Contest Problem Book II, AHSME 1961–65.
Vol. 25, The Contest Problem Book III, AHSME 1966–72.
Vol. 27, International Mathematical Olympiads, 1959–77
Vol. 29, The Contest Problem Book IV, AHSME 1973–82.
Vol. 31, International Mathematical Olympiads, 1978–85.
Vol. 33, USA Mathematical Olympiads, 1972–86.
Vol. 38, The Contest Problem Book V, AHSME 1983–88, AIME 1983–88.

Volume 38 also contains a guide to problem literature listing over 80 other books of interest to mathematical problem solvers. One series of pamphlets not listed there is *The Arbelos* by Samuel Greitzer, available through the American Mathematics Competitions office in Lincoln, NE 68501. Professor Greitzer was the first chair of the committee for the USA Mathematical Olympiad, and was instrumental in having the USA participate in the International Mathematical Olympiad. The six volumes in this series are of special interest to students preparing for free-response examinations in mathematics. They especially help to make USA students competitive in geometry within the international community.

Use of calculators

Permission to use calculators on the AHSME started with the 45th AHSME, the last one in this book. The use of calculators makes the AHSME accessible to more students. Because of the time limit, students who will solve all the problems on the test will not have much time for exploration on their calculators, just verification in most cases. Students not able to work all the problems on the AHSME can use their additional time to explore some of the questions with the aid of their calculator.

Calculator use, combined with the multiple-choice nature of the AHSME, prohibits the appearance of some problems on the test. For example, problem 6 on the 42nd AHSME,

$$\sqrt{x\sqrt{x\sqrt{x}}} = ?$$

could not have been used on the 45th AHSME or later since almost any positive $x \neq 1$ shows which is the correct choice. Also, problem 5 on the 41st AHSME which asks for the largest of

$$\sqrt{\sqrt[3]{5\cdot 6}},\ \sqrt{6\sqrt[3]{5}},\ \sqrt{5\sqrt[3]{6}},\ \sqrt[3]{5\sqrt{6}},\ \sqrt[3]{6\sqrt{5}}$$

would not be a valid mathematical question when calculators are permitted.

Generating the Examinations

The process of creating each AHSME takes almost two years. Every spring a call for problems is made to all those on any of the committees or the advisory panel of the American Mathematics Competitions, asking them to submit up to five great ideas for problems they might have spawned. (For the creation of the 40th through 45th AHSMEs there were 97 new problems submitted for the 40th, 97 for the 41st, 86 for the 42nd, 114 for the 43rd, 111 for the 44th and 82 for the 45th.)

During the summer following the call for problems, all submissions are sent to about two dozen panel and committee members who volunteer to spend about three months evaluating the problems for mathematical correctness, interest, newness and difficulty. This initial evaluation usually culls the number to about one and a half times the number that will be used on the examination.

Thereafter, through about four successive drafts distributed to various reviewers, additional criticism of the problems is aired, resulting in even fewer problems for each draft. Statements of the questions are polished to make them as unambiguous as possible, and alternative solutions are created. In fact, through the whole process the committee and reviewers probably spend more time generating and refining solutions than is actually spent on the problems.

During the review process, some of the best "distractors" (wrong answers)† to the AHSME problems are discovered, because reviewers are also subject to common mistakes.

Distractors thought to be pure traps of hurried misreading and not checks on incorrect mathematical thinking are avoided. For example, on problem 10 on the 42nd AHSME the distractor 7 was not used because it would have uncovered confusion between "the number of integers which are lengths of chords" and "the number of chords that have integer length." There was already enough other substance to this problem that additional complications need not be introduced.

After this lengthy review process involving several iterations and many volunteers is complete, the examination committee of nine members meets in the spring. This is the only face to face meeting, and in eight hours the examination is set. At this meeting, each question and its solution is reviewed in detail. At this stage relatively few changes are made to the

† Some participants may have a less polite term for these distractors – "Gotcha!" comes to mind.

questions and solutions because they have already been reviewed so thoroughly. Many problems can be approached in many ways and, therefore, have many alternative solutions. The official solutions for the competition must fit on ten pages for the solutions pamphlet, so a non-trivial amount of time is spent on deciding which of those alternatives to reserve for publication in this book. The last thing the committee decides is the numerical order of the problems for the test — easiest first, most challenging last. (To prove that the committee is not infallible, try correlating the order of the problems on the test with the percentage of students obtaining the correct solutions to the problems.)

Printing the examinations takes place during the summer, counting and prepackaging the following fall, and distribution of the AHSMEs to participating schools in the winter. All of this work, from the time the examination is set through the reporting of the results to the schools, is handled by the very efficient administrative staff of the American Mathematics Competitions headquartered on the campus of the University of Nebraska in Lincoln. For the AHSMEs in this book, that office was headed by Professor Walter Mientka, Executive Director of the American Mathematics Competitions. The examination-writing committees and all participants in these examinations are deeply indebted to Dr. Mientka and his extreme devotion to this work.

Reaction to these tests

Students or teachers who feel that a question was misleading are invited to write to the committee. Although no scores were changed for any of the examinations discussed in this book as the result of such correspondence, the committee considers each letter as a serious mathematical comment, makes a decision, and writes a polite reply trying to answer the objection.

Teachers and students also write unsolicited letters of congratulations on the work the committee has done in creating the tests. Some of these letters contain suggestions for alternative solutions to our problems. Occasionally these suggestions cover ground not explored during our lengthy review process, and some of these have been used to enhance solutions in this book. One of the most creative solutions, submitted by a high school problem group a couple weeks after the 43rd AHSME, appears in this book as the second solution to problem 20, the solution in which the star is decomposed into n congruent triangles which are moved to have a common vertex, showing that the angle measures $360°/n$.

Creating the Problems

One of the questions most frequently asked of AHSME and AIME committee members, is "How do you come up with ideas for new problems?" These tests are not the creation of one person or even a small group of people. Nine committee members and up to two dozen members of the advisory panel each submit several problems for the examination. All year long, these people are alert for new ideas for problems, and they send in the best of their findings when the call for problems is issued.

The pigeonhole principle shows that the expected number of problems per test is less than one per submitter. Therefore, these tests exhibit the best of the best.

There is more of a discussion of the AHSME problems in the section, "An Insider's Look at the AHSME Problems" which follows the solutions to the AHSME.

Acknowledgments

This volume presents a compilation of the labors of all those on the committees or advisory panel of the American Mathematics Competitions who contributed problems or were involved in the review steps in constructing the AHSME. These contributors include: Mangho Ahuja, Richard Anderson, Donald Bentley, George Berzsenyi, Janice Blasberg, Steven Blasberg, Bruce Bombacher, Kenneth Brown, Gail Burrill, Tom Butts, Mary Lou Derwent, David Drennan, Vladimir Drobot, Pam Drummond, Joseph Estephan, Paul Foerster, Larry Ford, Richard Gibbs, George Gilbert, Frank Hacker, David Hankin, John Haverhals, Bryan Hearsey, Gary Hendren, Gerald Heuer, Darrell Horwath, John Hoyt, Anne Hudson, Elgin Johnston, Irwin Kaufman, Joseph Kennedy, Clark Kimberling, Thomas Knapp, Joseph Konhauser, Steven Maurer, Eugene McGovern, David Mead, Walter Mientka, Rogers Newman, Richard Parris, Gregg Patruno, Stanley Rabinowitz, Harold Reiter, Ian Richards, Wayne Roberts, Mark Saul, Vincent Schielack, Leo Schneider, Terry Shell, Charlyn Shepherd, Fraser Simpson, Gordan Skukan, Alexander Soifer, Al Tinsley, Dan Ullman, Dorothy Wendt, Dan Wensing, Peter Yff, and Paul Zeitz.

The publications staff at the headquarters of the Mathematical Association of America, and especially Beverly Ruedi, provided great help to increase the aesthetic presentations of the problems and solutions in this book.

Sponsors

The American Mathematics Competitions, a program of the Mathematical Association of America, are currently also sponsored by the Society of Actuaries, Mu Alpha Theta, National Council of Teachers of Mathematics, Casualty Actuarial Society, American Statistical Society, American Mathematical Association of Two-Year Colleges, American Mathematical Society, American Society of Pension Actuaries, Consortium for Mathematics and its Applications, Pi Mu Epsilon, National Association of Mathematicians, and School Science and Mathematics Association.

Name

Beginning in 2000, the AHSME, which was designed for students in grades 9 through 12, will bifurcate into AMS→10 and AMC→12, designed for students in grades 9 and 10, and for students in grades 11 and 12, respectively.

AHSME RULES

- This is a thirty question multiple choice test. Each question is followed by answers marked A, B, C, D, and E. Only one of these is correct.
- SCORING RULES: You will receive 5 points for each correct answer, 2 points for each problem left unanswered, and 0 points for each incorrect answer.
- Solve the problem carefully. Note the scoring rules. To guess before eliminating 3 of the 5 choices will, on the average, lower your score.
- Scratch paper, graph paper, ruler, compass, protractor and eraser are permitted.

    Prior to 1994: Calculators and slide rules are not permitted.

    1994 to present: Any 'non-querty' calculator is permitted. No problems on the test will *require* the use of a calculator.
- Figures are not necessarily drawn to scale.
- When your proctor gives the signal, begin working the problems. You will have **90 MINUTES** working time for the test.

Computer scoring of the AHSME began in 1992. Other rules before 1992 concerned correctly and legibly completing the answer sheet. In 1992 these rules changed to directions about correctly encoding the computer-scanned answer sheet.

To insure integrity of the results of the AHSME, on the advice of lawyers every examination has, for a long time, contained these words at the bottom of the first page:

The results of this AHSME are used to identify students with unusual mathematical ability. To assure that this purpose is served, the Committee on the American Mathematics Competitions reserves the right to re-examine students before deciding whether to grant official status to individual or team scores. Re-examination will be requested when, after an inquiry, there is reasonable basis to believe that scores have been obtained by extremely lucky guessing or dishonesty. Official status will not be granted if a student or school does not agree to a requested re-examination. The committee also reserves the right to disqualify all scores from a school if it is determined that the required security procedures were not followed.

40 AHSME

1. $(-1)^{5^2} + 1^{2^5} =$

   (A) -7 **(B)** -2 **(C)** 0 **(D)** 1 **(E)** 57

2. $\sqrt{\frac{1}{9} + \frac{1}{16}} =$

   (A) $\frac{1}{5}$ **(B)** $\frac{1}{4}$ **(C)** $\frac{2}{7}$ **(D)** $\frac{5}{12}$ **(E)** $\frac{7}{12}$

3. A square is cut into three rectangles along two lines parallel to a side, as shown. If the perimeter of each of the three rectangles is 24, then the area of the original square is

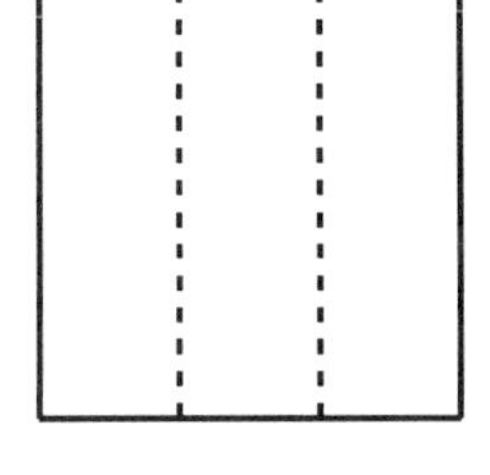

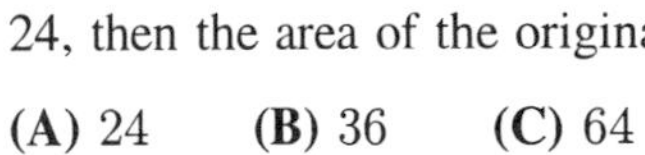

   (A) 24 **(B)** 36 **(C)** 64

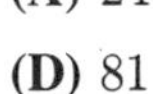

   (D) 81 **(E)** 96

4. In the figure, $ABCD$ is an isosceles trapezoid with side lengths $AD = BC = 5$, $AB = 4$, and $DC = 10$. The point C is on $\overline{DF}$ and B is the midpoint of hypotenuse $\overline{DE}$ in the right triangle DEF. Then $CF =$

   (A) 3.25 **(B)** 3.5 **(C)** 3.75

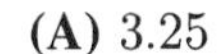

   (D) 4.0 **(E)** 4.25

5. Toothpicks of equal length are used to build a rectangular grid as shown. If the grid is 20 toothpicks high and 10 toothpicks wide, then the number of toothpicks used is

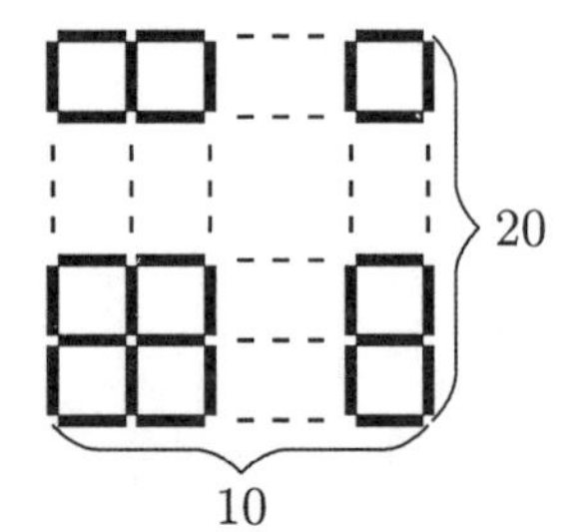

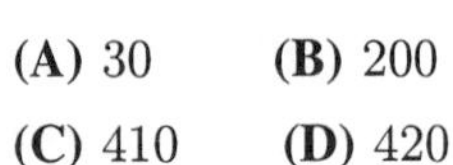

(A) 30 (B) 200

(C) 410 (D) 420

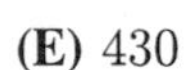

(E) 430

6. If $a, b > 0$ and the triangle in the first quadrant bounded by the coordinate axes and the graph of $ax + by = 6$ has area 6, then $ab =$

(A) 3 (B) 6 (C) 12 (D) 108 (E) 432

7. In $\triangle ABC$, $\angle A = 100°$, $\angle B = 50°$, $\angle C = 30°$, $\overline{AH}$ is an altitude, and $\overline{BM}$ is a median. Then $\angle MHC =$

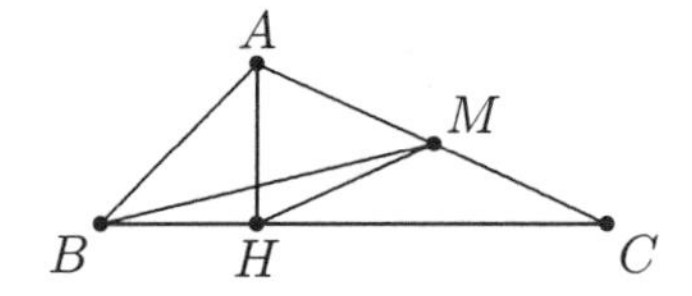

(A) 15° (B) 22.5°

(C) 30° (D) 40°

(E) 45°

8. For how many integers n between 1 and 100 does $x^2 + x - n$ factor into the product of two linear factors with integer coefficients?

(A) 0 (B) 1 (C) 2 (D) 9 (E) 10

9. Mr. and Mrs. Zeta want to name their baby Zeta so that its monogram (first, middle, and last initials) will be in alphabetical order with no letters repeated. How many such monograms are possible?

(A) 276 (B) 300 (C) 552 (D) 600 (E) 15600

10. Consider the sequence defined recursively by $u_1 = a$ (any positive number), and $u_{n+1} = -1/(u_n + 1)$, $n = 1, 2, 3, \ldots$. For which of the following values of n must $u_n = a$?

(A) 14 (B) 15 (C) 16 (D) 17 (E) 18

11. Let a, b, c and d be integers with $a < 2b$, $b < 3c$, and $c < 4d$. If $d < 100$, the largest possible value for a is

(A) 2367 **(B)** 2375 **(C)** 2391 **(D)** 2399 **(E)** 2400

12. The traffic on a certain east-west highway moves at a constant speed of 60 miles per hour in both directions. An eastbound driver passes 20 westbound vehicles in a five-minute interval. Assume vehicles in the westbound lane are equally spaced. Which of the following is closest to the number of westbound vehicles present in a 100-mile section of highway?

(A) 100 **(B)** 120 **(C)** 200 **(D)** 240 **(E)** 400

13. Two strips of width 1 overlap at an angle of α as shown. The area of the overlap (shown shaded) is

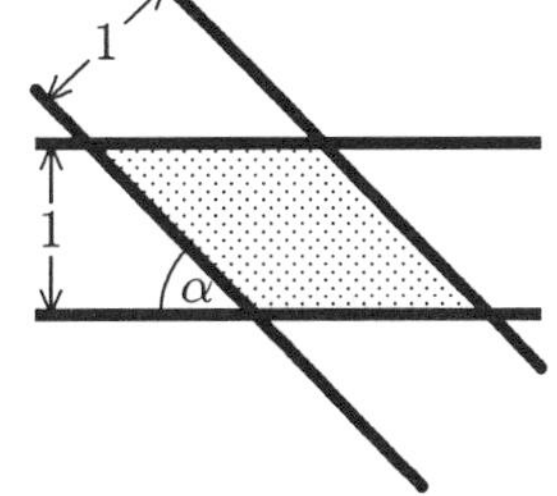

(A) $\sin\alpha$ **(B)** $\dfrac{1}{\sin\alpha}$

(C) $\dfrac{1}{1-\cos\alpha}$ **(D)** $\dfrac{1}{\sin^2\alpha}$

(E) $\dfrac{1}{(1-\cos\alpha)^2}$

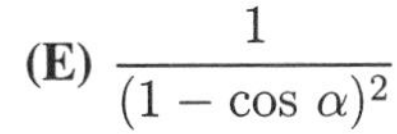

14. $\cot 10 + \tan 5 =$

(A) $\csc 5$ **(B)** $\csc 10$ **(C)** $\sec 5$

(D) $\sec 10$ **(E)** $\sin 15$

15. In $\triangle ABC$, $AB = 5$, $BC = 7$, $AC = 9$ and D is on $\overline{AC}$ with $BD = 5$. Find the ratio $AD : DC$.

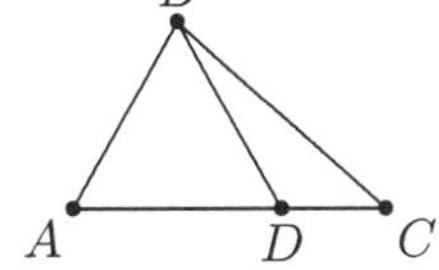

(A) $4:3$ **(B)** $7:5$ **(C)** $11:6$

(D) $13:5$ **(E)** $19:8$

16. A *lattice point* is a point in the plane with integer coordinates. How many lattice points are on the line segment whose endpoints are $(3, 17)$ and $(48, 281)$? (Include both endpoints of the segment in your count.)

(A) 2 **(B)** 4 **(C)** 6 **(D)** 16 **(E)** 46

17. The perimeter of an equilateral triangle exceeds the perimeter of a square by 1989 cm. The length of each side of the triangle exceeds the length of each side of the square by d cm. The square has perimeter greater than 0. How many positive integers are **not** possible values for d?

    (A) 0 **(B)** 9 **(C)** 221 **(D)** 663 **(E)** infinitely many

18. The set of all real numbers x for which

$$x+\sqrt{x^2+1}-\frac{1}{x+\sqrt{x^2+1}}$$

is a rational number is the set of all

(A) integers x **(B)** rational x **(C)** real x
(D) x for which $\sqrt{x^2+1}$ is rational
(E) x for which $x+\sqrt{x^2+1}$ is rational

19. A triangle is inscribed in a circle. The vertices of the triangle divide the circle into three arcs of lengths 3, 4, and 5. What is the area of the triangle?

    (A) 6 **(B)** $\frac{18}{\pi^2}$ **(C)** $\frac{9}{\pi^2}(\sqrt{3}-1)$
    (D) $\frac{9}{\pi^2}(\sqrt{3}+1)$ **(E)** $\frac{9}{\pi^2}(\sqrt{3}+3)$

20. Let x be a real number selected uniformly at random between 100 and 200. If $\lfloor\sqrt{x}\rfloor = 12$, find the probability that $\lfloor\sqrt{100x}\rfloor = 120$. ($\lfloor v\rfloor$ *means the greatest integer less than or equal to* v.)

    (A) $\frac{2}{25}$ **(B)** $\frac{241}{2500}$ **(C)** $\frac{1}{10}$ **(D)** $\frac{96}{625}$ **(E)** 1

21. A square flag has a red cross of uniform width with a blue square in the center on a white background as shown. (The cross is symmetric with respect to each of the diagonals of the square.) If the entire cross (both the red arms and the blue center) takes up 36% of the area of the flag, what percent of the area of the flag is blue?

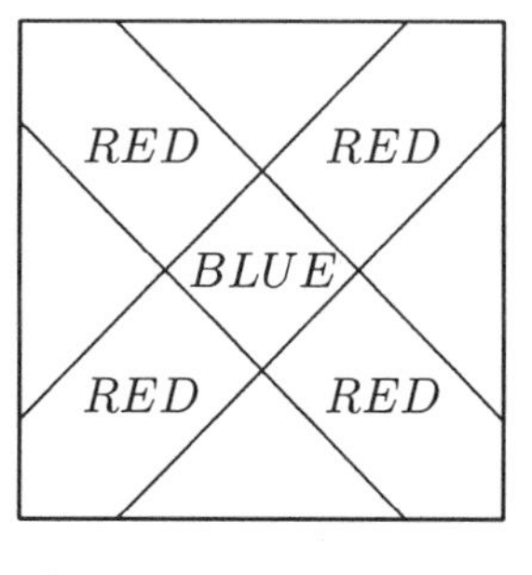

(A) .5 **(B)** 1 **(C)** 2 **(D)** 3 **(E)** 6

22. A child has a set of 96 distinct blocks. Each block is one of 2 materials (*plastic, wood*), 3 sizes (*small, medium, large*), 4 colors (*blue, green, red, yellow*), and 4 shapes (*circle, hexagon, square, triangle*). How many blocks in the set are different from the *"plastic medium red circle"* in exactly two ways? (The *"wood medium red square"* is such a block.)

(A) 29 **(B)** 39 **(C)** 48 **(D)** 56 **(E)** 62

23. A particle moves through the first quadrant as follows. During the first minute it moves from the origin to $(1,0)$. Thereafter, it continues to follow the directions indicated in the figure, going back and forth between the positive x and y axes, moving one unit of distance parallel to an axis in each minute. At which point will the particle be after exactly 1989 minutes?

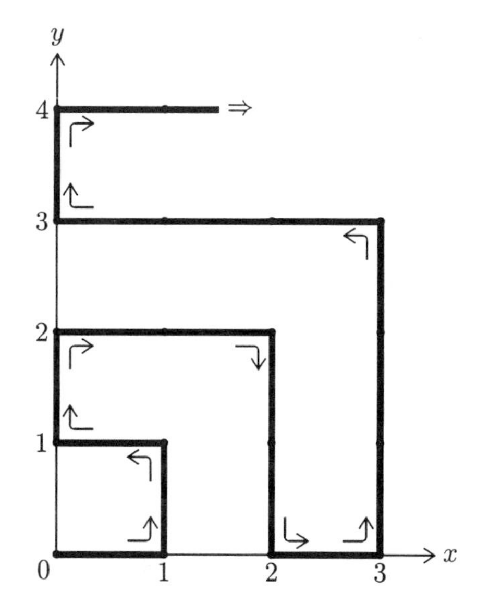

(A) (35,44)

(B) (36,45)

(C) (37,45)

(D) (44,35)

(E) (45,36)

24. Five people are sitting at a round table. Let $f \geq 0$ be the number of people sitting next to at least one female and $m \geq 0$ be the number of people sitting next to at least one male. The number of possible ordered pairs (f, m) is

    (A) 7 **(B)** 8 **(C)** 9 **(D)** 10 **(E)** 11

25. In a certain cross-country meet between two teams of five runners each, a runner who finishes in the nth position contributes n to his team's score. The team with the lower score wins. If there are no ties among the runners, how many different **winning** scores are possible?

    (A) 10 **(B)** 13 **(C)** 27 **(D)** 120 **(E)** 126

26. A regular octahedron is formed by joining the centers of adjoining faces of a cube. The ratio of the volume of the octahedron to the volume of the cube is

    (A) $\frac{\sqrt{3}}{12}$ **(B)** $\frac{\sqrt{6}}{16}$ **(C)** $\frac{1}{6}$ **(D)** $\frac{\sqrt{2}}{8}$ **(E)** $\frac{1}{4}$

27. Let n be a positive integer. If the equation $2x + 2y + z = n$ has 28 solutions in positive integers x, y and z, then n must be either

    (A) 14 or 15 **(B)** 15 or 16 **(C)** 16 or 17
    (D) 17 or 18 **(E)** 18 or 19

28. Find the sum of the roots of $\tan^2 x - 9\tan x + 1 = 0$ that are between $x = 0$ and $x = 2\pi$ radians.

    (A) $\frac{\pi}{2}$ **(B)** π **(C)** $\frac{3\pi}{2}$ **(D)** 3π **(E)** 4π

29. Find $\sum_{k=0}^{49}(-1)^k\binom{99}{2k}$, where $\binom{n}{j} = \frac{n!}{j!(n-j)!}$.

    (A) -2^{50} **(B)** -2^{49} **(C)** 0 **(D)** 2^{49} **(E)** 2^{50}

30. Suppose that 7 boys and 13 girls line up in a row. Let S be the number of places in the row where a boy and a girl are standing next to each other. For example, for the row $GBBGGGBGBGGGBGBGGBGG$

we have $S = 12$. The average value of S (if all possible orders of these 20 people are considered) is closest to

(A) 9 **(B)** 10 **(C)** 11 **(D)** 12 **(E)** 13

Some Comments on the Distractors

The table on the following page, which shows the distribution of responses for the top-scoring 0.47% of the students who participated in the 40th AHSME, showed the committee both expected and unexpected results. Most significant among these were:

- On problem 18, the number of students attracted by distractor (D) or (E) was over $3/4$ of the number who obtained the correct answer. This was somewhat expected since a shrewd guesser might expect a problem that looks complicated to have an answer that looks complicated.
- For problem 20, the committee expected (C) to be the most popular distractor since $12/120 = 1/10$. None of the distractors appears to have been exceedingly appealing to these top-scoring students. Hearsay evidence indicated that choice (C) was much more popular among students with lower scores. It is hard to explain why distractors (A) and (E) were both more popular among the honor roll students than was distractor (C).

Answers and Response Distribution

For the 40th AHSME, 401,889 copies of the examination were distributed to students at 6617 schools. There were 4 perfect papers. For the 1869 students named to the national Honor Roll for this examination, the following table lists the percent who gave the correct answer to each question. The percentages of responses to the other answers is also given.

	ANSWER				
#1	: (**C**) 97.29	(**A**) 0.06	(**B**) 0.11	(**D**) 0.06	(**E**) 0.00
#2	: (**D**) 99.23	(**A**) 0.00	(**B**) 0.00	(**C**) 0.00	(**E**) 0.39
#3	: (**D**) 97.12	(**A**) 2.32	(**B**) 2.32	(**C**) 0.06	(**E**) 0.33
#4	: (**D**) 88.16	(**A**) 0.11	(**B**) 0.11	(**C**) 0.22	(**E**) 0.22
#5	: (**E**) 93.25	(**A**) 0.06	(**B**) 0.33	(**C**) 0.28	(**D**) 0.39
#6	: (**A**) 87.60	(**B**) 2.21	(**C**) 4.65	(**D**) 0.33	(**E**) 0.50
#7	: (**C**) 83.23	(**A**) 0.17	(**B**) 0.39	(**D**) 0.28	(**E**) 0.33
#8	: (**D**) 89.65	(**A**) 0.17	(**B**) 0.17	(**C**) 0.06	(**E**) 2.93
#9	: (**B**) 72.83	(**A**) 1.44	(**C**) 0.66	(**D**) 2.93	(**E**) 5.15
#10	: (**C**) 69.56	(**A**) 0.28	(**B**) 0.89	(**D**) 1.22	(**E**) 0.39
#11	: (**A**) 80.52	(**B**) 3.82	(**C**) 5.70	(**D**) 8.63	(**E**) 0.33
#12	: (**C**) 82.01	(**A**) 1.00	(**B**) 0.66	(**D**) 0.61	(**E**) 9.13
#13	: (**B**) 80.13	(**A**) 3.04	(**C**) 0.11	(**D**) 1.72	(**E**) 0.22
#14	: (**B**) 61.43	(**A**) 1.11	(**C**) 0.50	(**D**) 0.44	(**E**) 0.11
#15	: (**E**) 38.85	(**A**) 0.33	(**B**) 0.94	(**C**) 0.94	(**D**) 2.60
#16	: (**B**) 81.41	(**A**) 2.21	(**C**) 1.05	(**D**) 1.33	(**E**) 0.17
#17	: (**D**) 68.79	(**A**) 0.17	(**B**) 0.17	(**C**) 0.33	(**E**) 6.36
#18	: (**B**) 39.62	(**A**) 0.66	(**C**) 5.31	(**D**) 10.85	(**E**) 20.14
#19	: (**E**) 25.84	(**A**) 1.27	(**B**) 0.44	(**C**) 0.28	(**D**) 1.38
#20	: (**B**) 33.04	(**A**) 9.24	(**C**) 4.37	(**D**) 0.39	(**E**) 7.08
#21	: (**C**) 34.37	(**A**) 0.39	(**B**) 1.38	(**D**) 1.38	(**E**) 1.88
#22	: (**A**) 62.81	(**B**) 0.55	(**C**) 0.44	(**D**) 1.00	(**E**) 2.49
#23	: (**D**) 38.85	(**A**) 1.05	(**B**) 1.05	(**C**) 0.61	(**E**) 2.60
#24	: (**B**) 36.58	(**A**) 4.59	(**C**) 1.33	(**D**) 1.33	(**E**) 0.28
#25	: (**B**) 29.11	(**A**) 0.44	(**C**) 2.38	(**D**) 1.33	(**E**) 5.59
#26	: (**C**) 32.98	(**A**) 0.11	(**B**) 0.17	(**D**) 0.72	(**E**) 2.27
#27	: (**D**) 12.51	(**A**) 0.94	(**B**) 0.94	(**C**) 1.27	(**E**) 1.33
#28	: (**D**) 9.57	(**A**) 1.22	(**B**) 1.22	(**C**) 0.89	(**E**) 2.82
#29	: (**B**) 5.76	(**A**) 0.89	(**C**) 2.99	(**D**) 1.72	(**E**) 0.11
#30	: (**A**) 3.82	(**B**) 0.83	(**C**) 1.38	(**D**) 0.39	(**E**) 0.50

41 AHSME

1. If $\dfrac{x/4}{2} = \dfrac{4}{x/2}$ then $x =$

   (A) $\pm 1/2$ **(B)** ± 1 **(C)** ± 2 **(D)** ± 4 **(E)** ± 8

2. $\left(\dfrac{1}{4}\right)^{-\frac{1}{4}} =$

   (A) -16 **(B)** $-\sqrt{2}$ **(C)** $-\dfrac{1}{16}$ **(D)** $\dfrac{1}{256}$ **(E)** $\sqrt{2}$

3. The consecutive angles of a trapezoid form an arithmetic sequence. If the smallest angle is 75°, then the largest angle is

   (A) 95° **(B)** 100° **(C)** 105° **(D)** 110° **(E)** 115°

4. Let $ABCD$ be a parallelogram with $\angle ABC = 120^\circ$, $AB = 16$ and $BC = 10$. Extend $\overline{CD}$ through D to E so that $DE = 4$. If $\overline{BE}$ intersects $\overline{AD}$ at F, then FD is closest to

   E 4 D C
   F
   10
   A 16 B

   (A) 1 **(B)** 2 **(C)** 3
   (D) 4 **(E)** 5

5. Which of these numbers is largest?

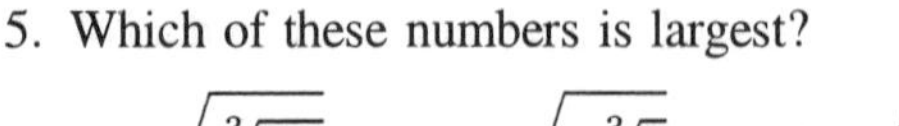

(A) $\sqrt{\sqrt[3]{5\cdot 6}}$ **(B)** $\sqrt{6\sqrt[3]{5}}$ **(C)** $\sqrt{5\sqrt[3]{6}}$

(D) $\sqrt[3]{5\sqrt{6}}$ **(E)** $\sqrt[3]{6\sqrt{5}}$

6. Points A and B are 5 units apart. How many lines in a given plane containing A and B are 2 units from A and 3 units from B?

(A) 0 **(B)** 1 **(C)** 2 **(D)** 3 **(E)** more than 3

7. A triangle with integral sides has perimeter 8. The area of the triangle is

(A) $2\sqrt{2}$ **(B)** $\frac{16}{9}\sqrt{3}$ **(C)** $2\sqrt{3}$ **(D)** 4 **(E)** $4\sqrt{2}$

8. The number of real solutions of the equation

$$|x-2|+|x-3|=1$$

is

(A) 0 **(B)** 1 **(C)** 2 **(D)** 3 **(E)** more than 3

9. Each edge of a cube is colored either red or black. Every face of the cube has at least one black edge. The smallest possible number of black edges is

(A) 2 **(B)** 3 **(C)** 4 **(D)** 5 **(E)** 6

10. An $11\times 11\times 11$ wooden cube is formed by gluing together 11^3 unit cubes. What is the greatest number of unit cubes that can be seen from a single point?

(A) 328 **(B)** 329 **(C)** 330 **(D)** 331 **(E)** 332

11. How many positive integers less than 50 have an odd number of positive integer divisors?

(A) 3 **(B)** 5 **(C)** 7 **(D)** 9 **(E)** 11

12. Let f be the function defined by $f(x) = ax^2 - \sqrt{2}$ for some positive a. If $f\big(f(\sqrt{2})\big) = -\sqrt{2}$ then $a =$

(A) $\frac{2-\sqrt{2}}{2}$ **(B)** $\frac{1}{2}$ **(C)** $2-\sqrt{2}$ **(D)** $\frac{\sqrt{2}}{2}$ **(E)** $\frac{2+\sqrt{2}}{2}$

13. If the following instructions are carried out by a computer, which value of X will be printed because of instruction 5?

```
1. START X AT 3 AND S AT 0.
2. INCREASE THE VALUE OF X BY 2.
3. INCREASE THE VALUE OF S BY THE VALUE OF X.
4. IF S IS AT LEAST 10000,
     THEN GO TO INSTRUCTION 5;
     OTHERWISE, GO TO INSTRUCTION 2
     AND PROCEED FROM THERE.
5. PRINT THE VALUE OF X.
6. STOP.
```

(A) 19 **(B)** 21 **(C)** 23 **(D)** 199 **(E)** 201

14. An acute isosceles triangle, ABC, is inscribed in a circle. Through B and C, tangents to the circle are drawn, meeting at point D. If $\angle ABC = \angle ACB = 2\angle D$ and x is the radian measure of $\angle A$, then $x =$

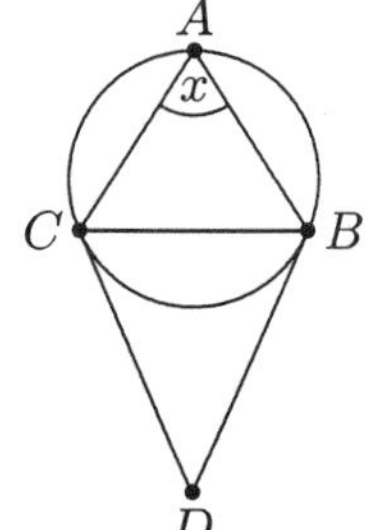

(A) $\frac{3}{7}\pi$ **(B)** $\frac{4}{9}\pi$ **(C)** $\frac{5}{11}\pi$

(D) $\frac{6}{13}\pi$ **(E)** $\frac{7}{15}\pi$

15. Four whole numbers, when added three at a time, give the sums 180, 197, 208 and 222. What is the largest of the four numbers?

(A) 77 **(B)** 83 **(C)** 89 **(D)** 95

(E) cannot be determined from the given information

16. At one of George Washington's parties, each man shook hands with everyone except his spouse, and no handshakes took place between

women. If 13 married couples attended, how many handshakes were there among these 26 people?

(A) 78 **(B)** 185 **(C)** 234 **(D)** 312 **(E)** 325

17. How many of the numbers, $100, 101, \ldots, 999$, have three different digits in increasing order or in decreasing order?

(A) 120 **(B)** 168 **(C)** 204 **(D)** 216 **(E)** 240

18. First a is chosen at random from the set $\{1, 2, 3, \ldots, 99, 100\}$, and then b is chosen at random from the same set. The probability that the integer $3^a + 7^b$ has units digit 8 is

(A) $\frac{1}{16}$ **(B)** $\frac{1}{8}$ **(C)** $\frac{3}{16}$ **(D)** $\frac{1}{5}$ **(E)** $\frac{1}{4}$

19. For how many integers N between 1 and 1990 is the improper fraction $\frac{N^2+7}{N+4}$ **not** in lowest terms?

(A) 0 **(B)** 86 **(C)** 90 **(D)** 104 **(E)** 105

20. In the figure, $ABCD$ is a quadrilateral with right angles at A and C. Points E and F are on $\overline{AC}$, and $\overline{DE}$ and $\overline{BF}$ are perpendicular to $\overline{AC}$. If $AE = 3$, $DE = 5$ and $CE = 7$, then $BF =$

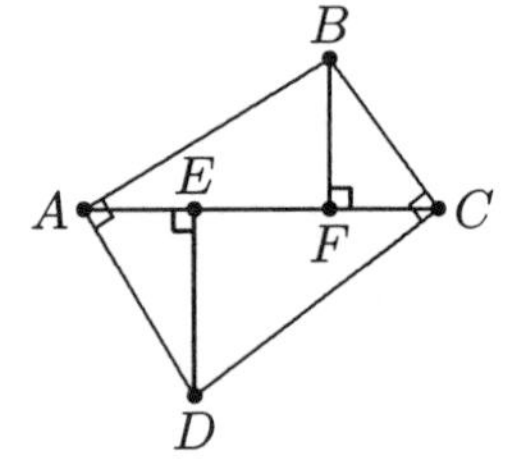

(A) 3.6 **(B)** 4 **(C)** 4.2
(D) 4.5 **(E)** 5

21. Consider a pyramid P-$ABCD$ whose base $ABCD$ is square and whose vertex P is equidistant from A, B, C and D. If $AB = 1$ and $\angle APB = 2\theta$ then the volume of the pyramid is

(A) $\frac{\sin\theta}{6}$ **(B)** $\frac{\cot\theta}{6}$ **(C)** $\frac{1}{6\sin\theta}$

(D) $\frac{1-\sin 2\theta}{6}$ **(E)** $\frac{\sqrt{\cos 2\theta}}{6\sin\theta}$

22. If the six solutions of $x^6 = -64$ are written in the form $a + bi$, where a and b are real, then the product of those solutions with $a > 0$ is

(A) -2 **(B)** 0 **(C)** $2i$ **(D)** 4 **(E)** 16

23. If $x, y > 0$, $\log_y x + \log_x y = \dfrac{10}{3}$ and $xy = 144$, then $\dfrac{x+y}{2} =$

(A) $12\sqrt{2}$ **(B)** $13\sqrt{3}$ **(C)** 24 **(D)** 30 **(E)** 36

24. All students at Adams High School and at Baker High School take a certain exam. The average scores for boys, for girls, and for boys and girls combined, at Adams HS and Baker HS are shown in the table, as is the average for boys at the two schools combined. What is the average score for the girls at the two schools combined?

	Adams	Baker	Adams & Baker
Boys :	71	81	79
Girls :	76	90	?
Boys & Girls :	74	84	

(A) 81 **(B)** 82 **(C)** 83 **(D)** 84 **(E)** 85

25. Nine congruent spheres are packed inside a unit cube in such a way that one of them has its center at the center of the cube and each of the others is tangent to the center sphere and to three faces of the cube. What is the radius of each sphere?

(A) $1 - \dfrac{\sqrt{3}}{2}$ **(B)** $\dfrac{2\sqrt{3}-3}{2}$ **(C)** $\dfrac{\sqrt{2}}{6}$

(D) $\dfrac{1}{4}$ **(E)** $\dfrac{\sqrt{3}(2-\sqrt{2})}{4}$

26. Ten people form a circle. Each picks a number and tells it to the two neighbors adjacent to him in the circle. Then each person computes and announces the average of the numbers of his two neighbors. The figure shows the average announced by each person (**not** the original number the person picked).

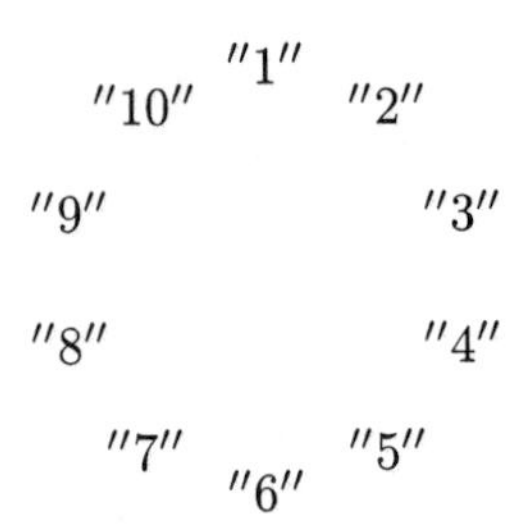

The number picked by the person who announced the average 6 was

(A) 1 **(B)** 5 **(C)** 6 **(D)** 10

(E) not uniquely determined from the given information

27. Which of these triples could <u>**not**</u> be the lengths of the three altitudes of a triangle?

(A) $1, \sqrt{3}, 2$ **(B)** $3, 4, 5$ **(C)** $5, 12, 13$

(D) $7, 8, \sqrt{113}$ **(E)** $8, 15, 17$

28. A quadrilateral that has consecutive sides of lengths 70, 90, 130 and 110 is inscribed in a circle and also has a circle inscribed in it. The point of tangency of the inscribed circle to the side of length 130 divides that side into segments of lengths x and y. Find $|x - y|$.

(A) 12 **(B)** 13 **(C)** 14 **(D)** 15 **(E)** 16

29. A subset of the integers $1, 2, \ldots, 100$ has the property that none of its members is 3 times another. What is the largest number of members such a subset can have?

(A) 50 **(B)** 66 **(C)** 67 **(D)** 76 **(E)** 78

30. If $R_n = \frac{1}{2}(a^n + b^n)$ where $a = 3 + 2\sqrt{2}$, $b = 3 - 2\sqrt{2}$, and $n = 0, 1, 2, \ldots$, then R_{12345} is an integer. Its units digit is

(A) 1 **(B)** 3 **(C)** 5 **(D)** 7 **(E)** 9

Some Comments on the Distractors

There were a few surprises for the AHSME committee when they studied the distribution of responses for the top-scoring 0.34% of the participants in the 41st AHSME:

- On problem 6, over 46% of the students chose distractor (A) or (B), so they either observed only the common tangent to the circles and not the other two exterior tangents, or vice versa. Well above an average amount of correspondence about this problem was initiated by participants who were probably not Honor Roll students. Their arguments that (E) was the correct answer indicated a lack of familiarity with the definition of the distance from a point to a line.
- The percent of students obtaining the correct answer to problem 29 indicates that this problem was much easier for the students than the committee had anticipated. Those who chose distractor (C) for this problem probably thought that no multiple of 3 could be in the subset.

Answers and Response Distribution

For the 41st AHSME, 394,214 copies of the examination were distributed to students at 6411 schools. There were 2 perfect papers. For the 1353 students named to the national Honor Roll for this examination, the following table lists the percent who gave the correct answer to each question. The percentages of responses to the other answers is also given.

	ANSWER				
#1	(E) 98.87	(A) 0.00	(B) 0.73	(C) 0.00	(D) 0.40
#2	(E) 99.19	(A) 0.00	(B) 0.08	(C) 0.00	(D) 0.24
#3	(C) 98.38	(A) 0.24	(B) 0.08	(D) 0.08	(E) 0.32
#4	(B) 87.16	(A) 0.32	(C) 2.67	(D) 0.97	(E) 0.24
#5	(B) 94.10	(A) 0.73	(C) 0.16	(D) 0.00	(E) 0.81
#6	(D) 51.78	(A) 23.10	(B) 23.10	(C) 2.34	(E) 5.41
#7	(A) 88.85	(B) 2.02	(C) 0.24	(D) 0.08	(E) 1.13
#8	(E) 81.83	(A) 0.32	(B) 1.29	(C) 11.55	(D) 3.07
#9	(B) 95.32	(A) 1.78	(C) 2.18	(D) 0.16	(E) 0.16
#10	(D) 88.61	(A) 0.24	(B) 0.24	(C) 1.53	(E) 2.75
#11	(C) 87.40	(A) 0.24	(B) 1.45	(D) 0.65	(E) 0.48
#12	(D) 94.83	(A) 0.00	(B) 0.00	(C) 0.32	(E) 0.40
#13	(E) 72.13	(A) 0.08	(B) 0.08	(C) 0.16	(D) 6.38
#14	(A) 41.52	(B) 0.81	(C) 0.16	(D) 0.24	(E) 0.97
#15	(C) 80.05	(A) 0.08	(B) 0.16	(D) 0.32	(E) 1.86
#16	(C) 88.13	(A) 0.65	(B) 0.08	(D) 4.85	(E) 0.48
#17	(C) 50.00	(A) 0.40	(B) 19.39	(D) 1.21	(E) 0.40
#18	(C) 50.65	(A) 1.94	(B) 14.05	(D) 0.73	(E) 3.47
#19	(B) 21.49	(A) 9.45	(C) 0.00	(D) 0.65	(E) 0.08
#20	(C) 21.24	(A) 0.40	(B) 1.53	(D) 1.37	(E) 4.85
#21	(E) 26.49	(A) 0.32	(B) 2.99	(C) 3.39	(D) 0.40
#22	(D) 26.66	(A) 2.42	(B) 2.42	(C) 0.65	(E) 3.07
#23	(B) 34.41	(A) 0.32	(C) 0.16	(D) 0.08	(E) 0.08
#24	(D) 54.68	(A) 2.18	(B) 2.18	(C) 0.97	(E) 0.65
#25	(B) 13.57	(A) 0.73	(C) 0.97	(D) 1.05	(E) 4.77
#26	(A) 29.08	(B) 0.81	(C) 1.21	(D) 0.32	(E) 3.31
#27	(C) 9.45	(A) 1.45	(B) 0.16	(D) 1.62	(E) 0.24
#28	(B) 1.45	(A) 0.16	(C) 0.48	(D) 0.40	(E) 0.24
#29	(D) 39.19	(A) 0.40	(B) 0.40	(C) 13.25	(E) 5.17
#30	(E) 3.39	(A) 1.94	(B) 3.88	(C) 0.32	(D) 1.13

42 AHSME

1. If for any three distinct numbers a, b and c we define $\boxed{a, b, c}$ by

$$\boxed{a, b, c} = \frac{c+a}{c-b},$$

then $\boxed{1, -2, -3} =$

(A) -2 **(B)** $-\frac{2}{5}$ **(C)** $-\frac{1}{4}$ **(D)** $\frac{2}{5}$ **(E)** 2

2. $|3 - \pi| =$

(A) $\frac{1}{7}$ **(B)** $.14$ **(C)** $3 - \pi$ **(D)** $3 + \pi$ **(E)** $\pi - 3$

3. $\left(4^{-1} - 3^{-1}\right)^{-1} =$

(A) -12 **(B)** -1 **(C)** $\frac{1}{12}$ **(D)** 1 **(E)** 12

4. Which of the following triangles cannot exist?

(A) An acute isosceles triangle

(B) An isosceles right triangle

(C) An obtuse right triangle

(D) A scalene right triangle

(E) A scalene obtuse triangle

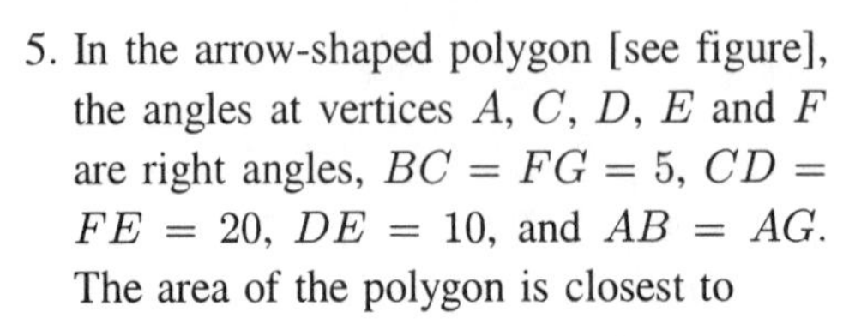

5. In the arrow-shaped polygon [see figure], the angles at vertices A, C, D, E and F are right angles, $BC = FG = 5$, $CD = FE = 20$, $DE = 10$, and $AB = AG$. The area of the polygon is closest to

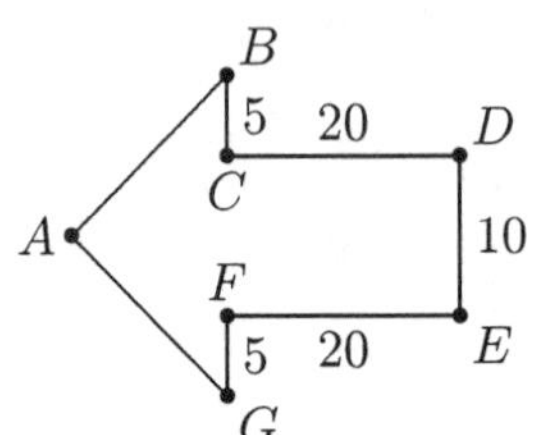

(A) 288 **(B)** 291 **(C)** 294

(D) 297 **(E)** 300

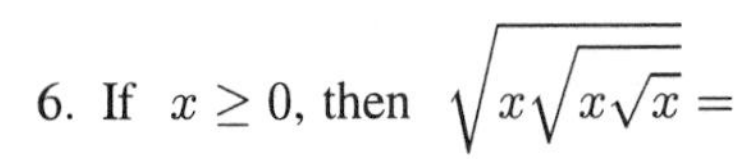

6. If $x \geq 0$, then $\sqrt{x\sqrt{x\sqrt{x}}} =$

(A) $x\sqrt{x}$ **(B)** $x\sqrt[4]{x}$ **(C)** $\sqrt[8]{x}$ **(D)** $\sqrt[8]{x^3}$ **(E)** $\sqrt[8]{x^7}$

7. If $x = a/b$, $a \neq b$ and $b \neq 0$, then $\dfrac{a+b}{a-b} =$

(A) $\dfrac{x}{x+1}$ **(B)** $\dfrac{x+1}{x-1}$ **(C)** 1 **(D)** $x - \dfrac{1}{x}$ **(E)** $x + \dfrac{1}{x}$

8. Liquid X does not mix with water. Unless obstructed, it spreads out on the surface of water to form a circular film 0.1 cm thick. A rectangular box measuring 6 cm by 3 cm by 12 cm is filled with liquid X. Its contents are poured onto a large body of water. What will be the radius, in centimeters, of the resulting circular film?

(A) $\dfrac{\sqrt{216}}{\pi}$ **(B)** $\sqrt{\dfrac{216}{\pi}}$ **(C)** $\sqrt{\dfrac{2160}{\pi}}$ **(D)** $\dfrac{216}{\pi}$ **(E)** $\dfrac{2160}{\pi}$

9. From time $t = 0$ to time $t = 1$ a population increased by $i\%$, and from time $t = 1$ to time $t = 2$ the population increased by $j\%$. Therefore, from time $t = 0$ to time $t = 2$ the population increased by

(A) $(i+j)\%$ **(B)** $ij\%$ **(C)** $(i+ij)\%$

(D) $\left(i + j + \dfrac{ij}{100}\right)\%$ **(E)** $\left(i + j + \dfrac{i+j}{100}\right)\%$

10. Point P is 9 units from the center of a circle of radius 15. How many different chords of the circle contain P and have integer lengths?

(A) 11 **(B)** 12 **(C)** 13 **(D)** 14 **(E)** 29

11. Jack and Jill run 10 kilometers. They start at the same point, run 5 kilometers up a hill, and return to the starting point by the same route. Jack has a 10-minute head start and runs at the rate of 15 km/hr uphill and 20 km/hr downhill. Jill runs 16 km/hr uphill and 22 km/hr downhill. How far from the top of the hill are they when they pass going in opposite directions?

(A) $\frac{5}{4}$ km **(B)** $\frac{35}{27}$ km **(C)** $\frac{27}{20}$ km **(D)** $\frac{7}{3}$ km **(E)** $\frac{28}{9}$ km

12. The measures (in degrees) of the interior angles of a convex hexagon form an arithmetic sequence of positive integers. Let m° be the measure of the largest interior angle of the hexagon. The largest possible value of m° is

(A) 165° **(B)** 167° **(C)** 170° **(D)** 175° **(E)** 179°

13. Horses X, Y and Z are entered in a three-horse race in which ties are not possible. If the odds against X winning are 3-to-1 and the odds against Y winning are 2-to-3, what are the odds against Z winning? (By *"odds against H winning are p-to-q"* we mean that the probability of H winning the race is $\frac{q}{p+q}$.)

(A) 3-to-20 **(B)** 5-to-6 **(C)** 8-to-5

(D) 17-to-3 **(E)** 20-to-3

14. If x is the cube of a positive integer and d is the number of positive integers that are divisors of x, then d could be

(A) 200 **(B)** 201 **(C)** 202 **(D)** 203 **(E)** 204

15. A circular table has exactly 60 chairs around it. There are N people seated at this table in such a way that the next person to be seated must sit next to someone. The smallest possible value of N is

(A) 15 **(B)** 20 **(C)** 30 **(D)** 40 **(E)** 58

16. One hundred students at Century High School participated in the **AHSME** last year, and their mean score was 100. The number of non-seniors taking the **AHSME** was 50% more than the number of

seniors, and the mean score of the seniors was 50% higher than that of the non-seniors. What was the mean score of the seniors?

(A) 100 **(B)** 112.5 **(C)** 120 **(D)** 125 **(E)** 150

17. A positive integer N is a *palindrome* if the integer obtained by reversing the sequence of digits of N is equal to N. The year 1991 is the only year in the current century with the following two properties:
    (a) It is a palindrome.
    (b) It factors as a product of a 2-digit prime palindrome and a 3-digit prime palindrome.

    How many years in the millenium between 1000 and 2000 (including the year 1991) have properties (a) and (b)?

    (A) 1 **(B)** 2 **(C)** 3 **(D)** 4 **(E)** 5

18. If S is the set of points z in the complex plane such that $(3+4i)z$ is a *real* number, then S is a

    (A) right triangle **(B)** circle **(C)** hyperbola
    (D) line **(E)** parabola

19. Triangle ABC has a right angle at C, $AC = 3$ and $BC = 4$. Triangle ABD has a right angle at A and $AD = 12$. Points C and D are on opposite sides of $\overline{AB}$. The line through D parallel to $\overline{AC}$ meets $\overline{CB}$ extended at E. If

$$\frac{DE}{DB} = \frac{m}{n},$$

where m and n are relatively prime positive integers, then $m + n =$

(A) 25 **(B)** 128 **(C)** 153 **(D)** 243 **(E)** 256

20. The sum of all real x such that

$$(2^x - 4)^3 + (4^x - 2)^3 = (4^x + 2^x - 6)^3$$

is

(A) $3/2$ **(B)** 2 **(C)** $5/2$ **(D)** 3 **(E)** $7/2$

21. If

$$f\left(\frac{x}{x-1}\right) = \frac{1}{x} \quad \text{for all } \ x \neq 0, 1$$

and $0 < \theta < \pi/2$, then $f\left(\sec^2 \theta\right) =$

(A) $\sin^2 \theta$ **(B)** $\cos^2 \theta$ **(C)** $\tan^2 \theta$ **(D)** $\cot^2 \theta$ **(E)** $\csc^2 \theta$

22. Two circles are externally tangent. Lines $\overline{PAB}$ and $\overline{PA'B'}$ are common tangents with A and A' on the smaller circle and B and B' on the larger circle. If $PA = AB = 4$, then the area of the smaller circle is

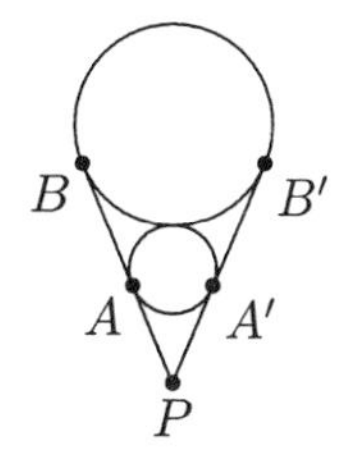

(A) 1.44π **(B)** 2π **(C)** 2.56π
(D) $\sqrt{8}\pi$ **(E)** 4π

23. If $ABCD$ is a 2×2 square, E is the midpoint of $\overline{AB}$, F is the midpoint of $\overline{BC}$, $\overline{AF}$ and $\overline{DE}$ intersect at I, and $\overline{BD}$ and $\overline{AF}$ intersect at H, then the area of quadrilateral $BEIH$ is

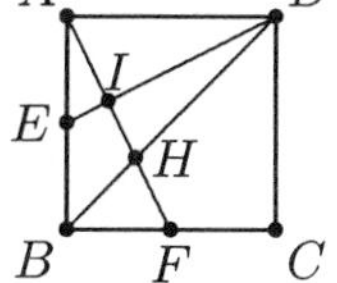

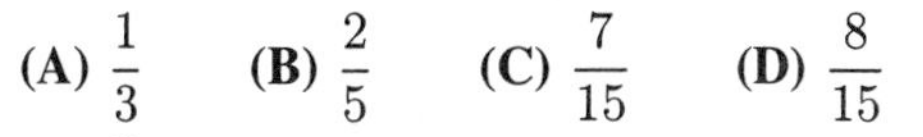

(A) $\frac{1}{3}$ **(B)** $\frac{2}{5}$ **(C)** $\frac{7}{15}$ **(D)** $\frac{8}{15}$
(E) $\frac{3}{5}$

24. The graph, G, of $y = \log_{10} x$ is rotated $90°$ counter-clockwise about the origin to obtain a new graph G'. Which of the following is an equation for G'?

(A) $y = \log_{10}\left(\frac{x+90}{9}\right)$ **(B)** $y = \log_x 10$ **(C)** $y = \frac{1}{x+1}$
(D) $y = 10^{-x}$ **(E)** $y = 10^x$

25. If $T_n = 1 + 2 + 3 + \cdots + n$ and

$$P_n = \frac{T_2}{T_2 - 1} \cdot \frac{T_3}{T_3 - 1} \cdot \frac{T_4}{T_4 - 1} \cdot \cdots \cdot \frac{T_n}{T_n - 1}$$

for $n = 2, 3, 4, \ldots$, then P_{1991} is closest to which of the following numbers?

(A) 2.0 **(B)** 2.3 **(C)** 2.6 **(D)** 2.9 **(E)** 3.2

26. An n-digit positive integer is *cute* if its n digits are an arrangement of the set $\{1, 2, \ldots, n\}$ and its first k digits form an integer that is divisible by k, for $k = 1, 2, \ldots, n$. For example, 321 is a cute 3-digit integer because 1 divides 3, 2 divides 32, and 3 divides 321. How many cute 6-digit integers are there?

(A) 0 **(B)** 1 **(C)** 2 **(D)** 3 **(E)** 4

27. If

$$x + \sqrt{x^2 - 1} + \frac{1}{x - \sqrt{x^2 - 1}} = 20$$

then

$$x^2 + \sqrt{x^4 - 1} + \frac{1}{x^2 + \sqrt{x^4 - 1}} =$$

(A) 5.05 **(B)** 20 **(C)** 51.005 **(D)** 61.25 **(E)** 400

28. Initially an urn contains 100 black marbles and 100 white marbles. Repeatedly, three marbles are removed from the urn and replaced from a pile outside the urn as follows:

MARBLES REMOVED	REPLACED WITH
3 black	1 black
2 black, 1 white	1 black, 1 white
1 black, 2 white	2 white
3 white	1 black, 1 white.

Which of the following sets of marbles could be the contents of the urn after repeated applications of this procedure?

(A) 2 black marbles **(B)** 2 white marbles **(C)** 1 black marble
(D) 1 black and 1 white marble **(E)** 1 white marble

29. Equilateral triangle ABC has been creased and folded so that vertex A now rests at A' on $\overline{BC}$ as shown. If $BA' = 1$ and $A'C = 2$ then the length of crease $\overline{PQ}$ is

(A) $\frac{8}{5}$ **(B)** $\frac{7}{20}\sqrt{21}$ **(C)** $\frac{1+\sqrt{5}}{2}$
(D) $\frac{13}{8}$ **(E)** $\sqrt{3}$

30. For any set S, let $|S|$ denote the number of elements in S, and let $n(S)$ be the number of subsets of S, including the empty set and the set S itself. If A, B and C are sets for which

$$n(A)+n(B)+n(C)=n(A\cup B\cup C) \quad \text{and} \quad |A|=|B|=100,$$

then what is the minimum possible value of $|A\cap B\cap C|$?

(A) 96 **(B)** 97 **(C)** 98 **(D)** 99 **(E)** 100

Some Comments on the Distractors

The table on the following page shows the distribution of responses for the top-scoring 1.22% of the participants in the 42nd AHSME. Problem 19 has, by far, the most interesting statistics. The percentage of students who chose distractor (A) is not only larger than the percentage of students who chose the correct answer, it is close to three times as large! There are ten fractions less than one whose numerator and denominator are relatively prime and sum to 25. Of these, only $12/13$ appears to have any connection to this problem. Those who chose this response must have guessed that DE was an integer. If that were the case, then $DE/DB = DE/13$, so 25 would be the maximum sum of the numerator and denominator of this fraction. If DE/DB were to equal $12/13$, then we would have $\triangle ABD \cong \triangle EBD$. The diagram does not make the triangles look congruent, but students know that AHSME diagrams are usually not accurate, lest students do geometry problems with a ruler.

Although problems 2 and 3 were both very easy, the statistics indicate less familiarity with the absolute value function than with negative powers and reciprocals.

Answers and Response Distribution

For the 42nd AHSME, 363,532 copies of the examination were distributed to students at 6120 schools. There were 3 perfect papers. For the 4418 students named to the national Honor Roll for this examination, the following table lists the percent who gave the correct answer to each question. The percentages of responses to the other answers is also given.

	ANSWER				
#1	: (E) 97.89	(A) 1.36	(B) 0.08	(C) 0.03	(D) 0.64
#2	: (E) 97.42	(A) 0.42	(B) 1.55	(C) 0.55	(D) 0.00
#3	: (A) 99.14	(B) 0.00	(C) 0.03	(D) 0.03	(E) 0.80
#4	: (C) 98.92	(A) 0.03	(B) 0.06	(D) 0.03	(E) 0.08
#5	: (E) 98.75	(A) 0.17	(B) 0.03	(C) 0.06	(D) 0.25
#6	: (E) 96.14	(A) 0.11	(B) 0.31	(C) 1.83	(D) 0.36
#7	: (B) 97.48	(A) 0.11	(C) 0.03	(D) 0.03	(E) 0.00
#8	: (C) 98.14	(A) 0.00	(B) 0.67	(D) 0.00	(E) 0.03
#9	: (D) 92.07	(A) 0.58	(B) 0.58	(C) 0.78	(E) 1.05
#10	: (B) 30.29	(A) 2.27	(C) 3.25	(D) 1.41	(E) 0.89
#11	: (B) 69.04	(A) 0.33	(C) 0.80	(D) 0.03	(E) 0.44
#12	: (D) 42.61	(A) 0.36	(B) 0.36	(C) 2.66	(E) 4.91
#13	: (D) 92.21	(A) 0.00	(B) 0.00	(C) 0.00	(E) 0.61
#14	: (C) 21.36	(A) 1.78	(B) 1.69	(D) 1.05	(E) 2.27
#15	: (B) 88.68	(A) 0.61	(C) 1.75	(D) 0.55	(E) 0.31
#16	: (D) 87.85	(A) 1.14	(B) 1.14	(C) 0.83	(E) 1.11
#17	: (D) 50.01	(A) 0.72	(B) 0.72	(C) 5.83	(E) 6.91
#18	: (D) 44.63	(A) 0.19	(B) 0.19	(C) 0.33	(E) 0.53
#19	: (B) 8.32	(A) 23.99	(C) 1.39	(D) 0.08	(E) 0.53
#20	: (E) 12.12	(A) 0.53	(B) 1.44	(C) 1.28	(D) 0.83
#21	: (A) 44.22	(B) 1.91	(C) 0.69	(D) 0.78	(E) 5.33
#22	: (B) 14.06	(A) 0.25	(C) 0.42	(D) 0.25	(E) 1.00
#23	: (C) 11.98	(A) 1.00	(B) 0.80	(D) 0.64	(E) 0.44
#24	: (D) 59.31	(A) 0.28	(B) 0.28	(C) 0.11	(E) 2.05
#25	: (D) 9.21	(A) 1.58	(B) 1.58	(C) 2.55	(E) 3.11
#26	: (C) 9.18	(A) 2.39	(B) 1.89	(D) 0.31	(E) 0.94
#27	: (C) 12.43	(A) 0.25	(B) 0.33	(D) 0.14	(E) 0.25
#28	: (B) 26.99	(A) 0.80	(C) 1.94	(D) 4.11	(E) 0.86
#29	: (B) 2.16	(A) 0.17	(C) 0.22	(D) 0.19	(E) 0.55
#30	: (B) 0.89	(A) 0.17	(C) 1.11	(D) 0.89	(E) 0.72

43 AHSME

1. If $3(4x + 5\pi) = P$, then $6(8x + 10\pi) =$

   (A) $2P$ **(B)** $4P$ **(C)** $6P$ **(D)** $8P$ **(E)** $18P$

2. An urn is filled with coins and beads, all of which are either silver or gold. Twenty percent of the objects in the urn are beads. Forty percent of the coins in the urn are silver. What percent of the objects in the urn are gold coins?

   (A) 40% **(B)** 48% **(C)** 52% **(D)** 60% **(E)** 80%

3. If $m > 0$ and the points $(m, 3)$ and $(1, m)$ lie on a line with slope m, then $m =$

   (A) 1 **(B)** $\sqrt{2}$ **(C)** $\sqrt{3}$ **(D)** 2 **(E)** $\sqrt{5}$

4. If a, b and c are positive integers and a and b are odd, then $3^a + (b-1)^2 c$ is

   (A) odd for all choices of c **(B)** even for all choices of c
   (C) odd if c is even; even if c is odd
   (D) odd if c is odd; even if c is even
   (E) odd if c is not a multiple of 3; even if c is a multiple of 3

5. $6^6 + 6^6 + 6^6 + 6^6 + 6^6 + 6^6 =$

   (A) 6^6 **(B)** 6^7 **(C)** 36^6 **(D)** 6^{36} **(E)** 36^{36}

6. If $x > y > 0$, then $\dfrac{x^y y^x}{y^y x^x} =$

(A) $(x-y)^{y/x}$ **(B)** $\left(\dfrac{x}{y}\right)^{x-y}$ **(C)** 1 **(D)** $\left(\dfrac{x}{y}\right)^{y-x}$

(E) $(x-y)^{x/y}$

7. The ratio of w to x is $4:3$, of y to z is $3:2$ and of z to x is $1:6$. What is the ratio of w to y?

(A) $1:3$ **(B)** $16:3$ **(C)** $20:3$ **(D)** $27:4$ **(E)** $12:1$

8. A square floor is tiled with congruent square tiles. The tiles on the two diagonals of the floor are black. The rest of the tiles are white. If there are 101 black tiles, then the total number of tiles is

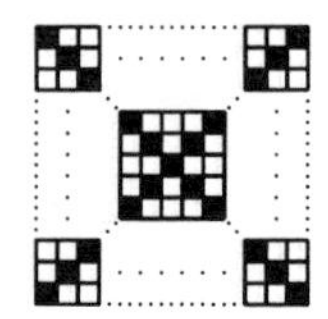

(A) 121 **(B)** 625 **(C)** 676 **(D)** 2500 **(E)** 2601

9. Five equilateral triangles, each with side $2\sqrt{3}$, are arranged so they are all on the same side of a line containing one side of each. Along this line, the midpoint of the base of one triangle is a vertex of the next. The area of the region of the plane that is covered by the union of the five triangular regions is

(A) 10 **(B)** 12 **(C)** 15 **(D)** $10\sqrt{3}$ **(E)** $12\sqrt{3}$

10. The number of positive integers k for which the equation

$$kx - 12 = 3k$$

has an integer solution for x is

(A) 3 **(B)** 4 **(C)** 5 **(D)** 6 **(E)** 7

11. The ratio of the radii of two concentric circles is $1:3$. If $\overline{AC}$ is a diameter of the larger circle, $\overline{BC}$ is a chord of the larger circle that is tangent to the smaller circle, and $AB = 12$, then the radius of the larger circle is

(A) 13 **(B)** 18 **(C)** 21 **(D)** 24 **(E)** 26

12. Let $y = mx + b$ be the image when the line $x - 3y + 11 = 0$ is reflected across the x-axis. The value of $m + b$ is

(A) -6 **(B)** -5 **(C)** -4 **(D)** -3 **(E)** -2

13. How many pairs of positive integers (a, b) with $a + b \leq 100$ satisfy the equation

$$\frac{a + b^{-1}}{a^{-1} + b} = 13?$$

(A) 1 **(B)** 5 **(C)** 7 **(D)** 9 **(E)** 13

14. Which of the following equations have the same graph?

I. $y = x - 2$ II. $y = \dfrac{x^2 - 4}{x + 2}$ III. $(x + 2)y = x^2 - 4$

(A) I and II only **(B)** I and III only

(C) II and III only **(D)** I, II and III

(E) None. All the equations have different graphs

15. Let $i = \sqrt{-1}$. Define a sequence of complex numbers by

$$z_1 = 0, \quad z_{n+1} = z_n^2 + i \text{ for } n \geq 1.$$

In the complex plane, how far from the origin is z_{111}?

(A) 1 **(B)** $\sqrt{2}$ **(C)** $\sqrt{3}$ **(D)** $\sqrt{110}$ **(E)** $\sqrt{2^{55}}$

16. If

$$\frac{y}{x - z} = \frac{x + y}{z} = \frac{x}{y}$$

for three positive numbers x, y and z, all different, then $\dfrac{x}{y} =$

(A) $\dfrac{1}{2}$ **(B)** $\dfrac{3}{5}$ **(C)** $\dfrac{2}{3}$ **(D)** $\dfrac{5}{3}$ **(E)** 2

17. The two-digit integers from 19 to 92 are written consecutively to form the large integer

$$N = 19202122\ldots909192.$$

If 3^k is the highest power of 3 that is a factor of N, then $k =$

(A) 0 **(B)** 1 **(C)** 2 **(D)** 3 **(E)** more than 3

18. The increasing sequence of positive integers $a_1, a_2, a_3, \ldots$ has the property that

$$a_{n+2} = a_n + a_{n+1} \quad \text{for all} \quad n \geq 1.$$

If $a_7 = 120$, then a_8 is

(A) 128 **(B)** 168 **(C)** 193 **(D)** 194 **(E)** 210

19. For each vertex of a solid cube, consider the tetrahedron determined by the vertex and the midpoints of the three edges that meet at that vertex. The portion of the cube that remains when these eight tetrahedra are cut away is called a *cuboctahedron*. The ratio of the volume of the cuboctahedron to the volume of the original cube is closest to which of these?

(A) 75% **(B)** 78% **(C)** 81% **(D)** 84% **(E)** 87%

20. Part of an "n-pointed regular star" is shown. It is a simple closed polygon in which all $2n$ edges are congruent, angles A_1, A_2, …, A_n are congruent and angles B_1, B_2, …, B_n are congruent. If the acute angle at A_1 is $10°$ less than the acute angle at B_1, then $n =$

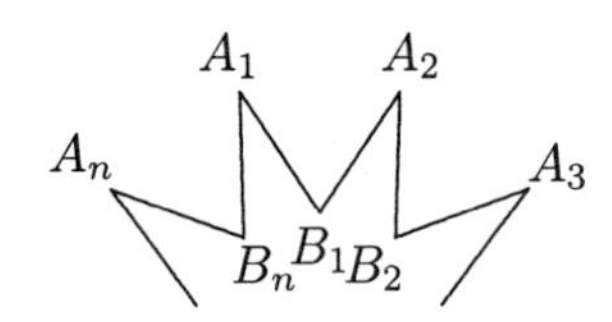

(A) 12 **(B)** 18 **(C)** 24 **(D)** 36 **(E)** 60

21. For a finite sequence $A = (a_1, a_2, \ldots, a_n)$ of numbers, the *Cesàro sum* of A is defined to be

$$\frac{S_1 + S_2 + \cdots + S_n}{n},$$

where

$$S_k = a_1 + a_2 + a_3 + \cdots + a_k \quad (1 \leq k \leq n).$$

If the Cesàro sum of the 99-term sequence $(a_1, a_2, \ldots, a_{99})$ is 1000, what is the Cesàro sum of the 100-term sequence $(1, a_1, a_2, \ldots, a_{99})$?

(A) 991 **(B)** 999 **(C)** 1000 **(D)** 1001 **(E)** 1009

22. Ten points are selected on the positive x-axis, $\mathbf{X}^+$, and five points are selected on the positive y-axis, $\mathbf{Y}^+$. The fifty segments connecting the ten selected points on $\mathbf{X}^+$ to the five selected points on $\mathbf{Y}^+$ are drawn. What is the maximum possible number of points of intersection of these fifty segments that could lie in the interior of the first quadrant?

    (A) 250 **(B)** 450 **(C)** 500 **(D)** 1250 **(E)** 2500

23. What is the size of the largest subset, S, of $\{1, 2, 3, \ldots, 50\}$ such that no pair of distinct elements of S has a sum divisible by 7?

    (A) 6 **(B)** 7 **(C)** 14 **(D)** 22 **(E)** 23

24. Let $ABCD$ be a parallelogram of area 10 with $AB = 3$ and $BC = 5$. Locate E, F and G on segments $\overline{AB}$, $\overline{BC}$ and $\overline{AD}$, respectively, with $AE = BF = AG = 2$. Let the line through G parallel to $\overline{EF}$ intersect $\overline{CD}$ at H. The area of the quadrilateral $EFHG$ is

    (A) 4 **(B)** 4.5 **(C)** 5 **(D)** 5.5 **(E)** 6

25. In triangle ABC, $\angle ABC = 120°$, $AB = 3$ and $BC = 4$. If perpendiculars constructed to $\overline{AB}$ at A and to $\overline{BC}$ at C meet at D, then $CD =$

    (A) 3 **(B)** $\dfrac{8}{\sqrt{3}}$ **(C)** 5 **(D)** $\dfrac{11}{2}$ **(E)** $\dfrac{10}{\sqrt{3}}$

26. Semicircle $\widehat{AB}$ has center C and radius 1. Point D is on $\widehat{AB}$ and $\overline{CD} \perp \overline{AB}$. Extend $\overline{BD}$ and $\overline{AD}$ to E and F, respectively, so that circular arcs $\widehat{AE}$ and BF have B and A as their respective centers. Circular arc EF has center D. The area of the shaded "smile", $AEFBDA$, is

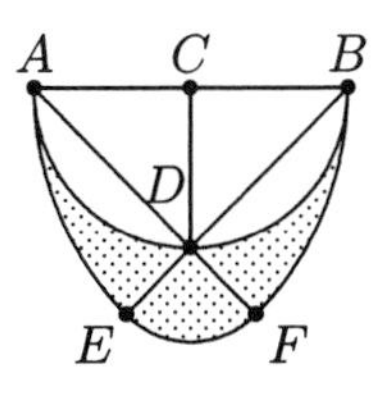

    (A) $\left(2 - \sqrt{2}\right)\pi$ **(B)** $2\pi - \pi\sqrt{2} - 1$ **(C)** $\left(1 - \dfrac{\sqrt{2}}{2}\right)\pi$

    (D) $\dfrac{5\pi}{2} - \pi\sqrt{2} - 1$ **(E)** $\left(3 - 2\sqrt{2}\right)\pi$

27. A circle of radius r has chords $\overline{AB}$ of length 10 and $\overline{CD}$ of length 7. When $\overline{AB}$ and $\overline{CD}$ are extended through B and C, respectively, they intersect at P, which is outside the circle. If $\angle APD = 60°$ and $BP = 8$, then $r^2 =$

    (A) 70 **(B)** 71 **(C)** 72 **(D)** 73 **(E)** 74

28. Let $i = \sqrt{-1}$. The product of the real parts of the roots of $z^2 - z = 5 - 5i$ is

    (A) -25 **(B)** -6 **(C)** -5 **(D)** $\frac{1}{4}$ **(E)** 25

29. An "unfair" coin has a 2/3 probability of turning up heads. If this coin is tossed 50 times, what is the probability that the total number of heads is even?

    (A) $25\left(\frac{2}{3}\right)^{50}$ **(B)** $\frac{1}{2}\left(1 - \frac{1}{3^{50}}\right)$ **(C)** $\frac{1}{2}$

    (D) $\frac{1}{2}\left(1 + \frac{1}{3^{50}}\right)$ **(E)** $\frac{2}{3}$

30. Let $ABCD$ be an isosceles trapezoid with bases $AB = 92$ and $CD = 19$. Suppose $AD = BC = x$ and a circle with center on $\overline{AB}$ is tangent to segments $\overline{AD}$ and $\overline{BC}$. If m is the smallest possible value of x, then $m^2 =$

    (A) 1369 **(B)** 1679 **(C)** 1748 **(D)** 2109 **(E)** 8825

Some Comments on the Distractors

The table on the following page containing data on the 1.52% top-scoring participants in the 43rd AHSME reveals that in two problems, 14 and 28, there was a distractor much more attractive to the participants than was the correct answer:

- Many in the elite group represented by this data probably were familiar with calculus in which it is emphasized that $y = x - 2$ and $y = (x^2-4)/(x+2)$ have different graphs. Therefore, in problem 14 choice (D) was not the popular choice the committee expected. However, over 1/3 of the students in this group failed to consider the vertical line which is part of the graph of $(x+2)y = x^2 - 4$, and they therefore chose distractor (C).

- Only about 10% answered problem 28, but those who did chose the distractor (D) more than the correct answer. Distractor (D) is the only non-integer among the choices. The only explanation that occurred to the committee is that the denominator of that fraction is 4, and one might expect both solutions to have a denominator of 2 when one considers the quadratic formula.

Answers and Response Distribution

For the 43rd AHSME, 355,319 copies of the examination were distributed to students at 5920 schools. There were 4 perfect papers. For the 5417 students named to the national Honor Roll for this examination, the following table lists the percent who gave the correct answer to each question. The percentages of responses to the other answers is also given.

	ANSWER				
#1	: (B) 99.00	(A) 0.18	(C) 0.31	(D) 0.37	(E) 0.04
#2	: (B) 99.55	(A) 0.29	(C) 0.06	(D) 0.00	(E) 0.00
#3	: (B) 96.32	(A) 2.23	(C) 0.14	(D) 0.55	(E) 0.02
#4	: (C) 98.36	(A) 0.33	(B) 0.08	(D) 0.27	(E) 0.12
#5	: (A) 95.60	(B) 0.63	(C) 2.93	(D) 0.41	(E) 0.04
#6	: (D) 97.79	(A) 0.88	(B) 0.88	(C) 0.08	(E) 0.00
#7	: (B) 98.96	(A) 0.27	(C) 0.02	(D) 0.10	(E) 0.49
#8	: (E) 93.31	(A) 0.04	(B) 0.14	(C) 0.55	(D) 3.62
#9	: (E) 95.87	(A) 0.02	(B) 0.45	(C) 0.02	(D) 0.31
#10	: (D) 93.76	(A) 0.61	(B) 0.61	(C) 2.03	(E) 0.47
#11	: (B) 78.28	(A) 0.14	(C) 0.14	(D) 0.88	(E) 0.10
#12	: (C) 96.11	(A) 0.04	(B) 0.02	(D) 0.22	(E) 0.04
#13	: (C) 74.52	(A) 1.00	(B) 0.16	(D) 0.14	(E) 0.10
#14	: (E) 25.26	(A) 0.43	(B) 13.16	(C) 37.33	(D) 16.84
#15	: (B) 35.39	(A) 3.99	(C) 0.10	(D) 0.51	(E) 1.86
#16	: (E) 68.13	(A) 0.88	(B) 0.20	(C) 0.12	(D) 0.55
#17	: (B) 46.03	(A) 5.30	(C) 1.08	(D) 0.84	(E) 1.00
#18	: (D) 58.84	(A) 0.25	(B) 0.25	(C) 1.84	(E) 0.65
#19	: (D) 20.22	(A) 0.80	(B) 0.80	(C) 1.02	(E) 2.87
#20	: (D) 24.85	(A) 1.11	(B) 1.11	(C) 0.10	(E) 0.14
#21	: (A) 41.45	(B) 0.39	(C) 1.92	(D) 3.17	(E) 0.49
#22	: (B) 21.72	(A) 0.51	(C) 0.86	(D) 1.33	(E) 0.29
#23	: (E) 22.04	(A) 1.08	(B) 1.31	(C) 1.11	(D) 6.24
#24	: (C) 13.79	(A) 0.57	(B) 0.47	(D) 0.49	(E) 0.72
#25	: (E) 14.37	(A) 1.00	(B) 0.72	(C) 0.65	(D) 0.84
#26	: (B) 20.77	(A) 1.76	(C) 0.31	(D) 1.78	(E) 0.10
#27	: (D) 1.11	(A) 0.06	(B) 0.06	(C) 0.41	(E) 0.02
#28	: (B) 3.75	(A) 0.35	(C) 1.80	(D) 4.15	(E) 0.20
#29	: (D) 4.13	(A) 1.37	(B) 1.37	(C) 2.41	(E) 0.63
#30	: (B) 1.29	(A) 0.49	(C) 0.29	(D) 0.33	(E) 0.14

44 AHSME

1. For integers a, b and c, define $\boxed{a, b, c}$ to mean $a^b - b^c + c^a$. Then $\boxed{1, -1, 2}$ equals

   (A) -4 **(B)** -2 **(C)** 0 **(D)** 2 **(E)** 4

2. In $\triangle ABC$, $\angle A = 55°$, $\angle C = 75°$, D is on side $\overline{AB}$ and E is on side $\overline{BC}$. If $DB = BE$, then $\angle BED =$

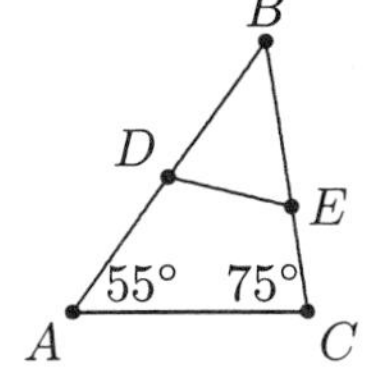

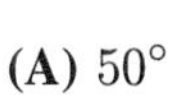

   (A) $50°$ **(B)** $55°$ **(C)** $60°$

   (D) $65°$ **(E)** $70°$

3. $\dfrac{15^{30}}{45^{15}} =$

   (A) $\left(\dfrac{1}{3}\right)^{15}$ **(B)** $\left(\dfrac{1}{3}\right)^{2}$ **(C)** 1 **(D)** 3^{15} **(E)** 5^{15}

4. Define the operation "$\circ$" by $x \circ y = 4x - 3y + xy$, for all real numbers x and y. For how many real numbers y does $3 \circ y = 12$?

   (A) 0 **(B)** 1 **(C)** 3 **(D)** 4 **(E)** more than 4

5. Last year a bicycle cost \$160 and a cycling helmet cost \$40. This year the cost of the bicycle increased by 5%, and the cost of the helmet

increased by 10%. The percent increase in the combined cost of the bicycle and the helmet is

(A) 6% **(B)** 7% **(C)** 7.5% **(D)** 8% **(E)** 15%

6. $\sqrt{\dfrac{8^{10}+4^{10}}{8^4+4^{11}}} =$

(A) $\sqrt{2}$ **(B)** 16 **(C)** 32 **(D)** $12^{2/3}$ **(E)** 512.5

7. The symbol R_k stands for an integer whose base-ten representation is a sequence of k ones. For example, $R_3 = 111$, $R_5 = 11111$, etc. When R_{24} is divided by R_4, the quotient $Q = R_{24}/R_4$ is an integer whose base-ten representation is a sequence containing only ones and zeros. The number of zeros in Q is

(A) 10 **(B)** 11 **(C)** 12 **(D)** 13 **(E)** 15

8. Let C_1 and C_2 be circles of radius 1 that are in the same plane and tangent to each other. How many circles of radius 3 are in this plane and tangent to both C_1 and C_2?

(A) 2 **(B)** 4 **(C)** 5 **(D)** 6 **(E)** 8

9. Country $\mathcal{A}$ has $c\%$ of the world's population and owns $d\%$ of the world's wealth. Country $\mathcal{B}$ has $e\%$ of the world's population and $f\%$ of its wealth. Assume that the citizens of $\mathcal{A}$ share the wealth of $\mathcal{A}$ equally, and assume that those of $\mathcal{B}$ share the wealth of $\mathcal{B}$ equally. Find the ratio of the wealth of a citizen of $\mathcal{A}$ to the wealth of a citizen of $\mathcal{B}$.

(A) $\dfrac{cd}{ef}$ **(B)** $\dfrac{ce}{df}$ **(C)** $\dfrac{cf}{de}$ **(D)** $\dfrac{de}{cf}$ **(E)** $\dfrac{df}{ce}$

10. Let r be the number that results when both the base and the exponent of a^b are tripled, where $a, b > 0$. If r equals the product of a^b and x^b where $x > 0$, then $x =$

(A) 3 **(B)** $3a^2$ **(C)** $27a^2$ **(D)** $2a^{3b}$ **(E)** $3a^{2b}$

11. If $\log_2(\log_2(\log_2(x))) = 2$, then how many digits are in the base-ten representation for x?

(A) 5 **(B)** 7 **(C)** 9 **(D)** 11 **(E)** 13

12. If $f(2x) = \dfrac{2}{2+x}$ for all $x > 0$, then $2f(x) =$

(A) $\dfrac{2}{1+x}$ **(B)** $\dfrac{2}{2+x}$ **(C)** $\dfrac{4}{1+x}$

(D) $\dfrac{4}{2+x}$ **(E)** $\dfrac{8}{4+x}$

13. A square of perimeter 20 is inscribed in a square of perimeter 28. What is the greatest distance between a vertex of the inner square and a vertex of the outer square?

(A) $\sqrt{58}$ **(B)** $\dfrac{7\sqrt{5}}{2}$ **(C)** 8 **(D)** $\sqrt{65}$ **(E)** $5\sqrt{3}$

14. The convex pentagon $ABCDE$ has $\angle A = \angle B = 120^\circ$, $EA = AB = BC = 2$ and $CD = DE = 4$. What is the area of $ABCDE$?

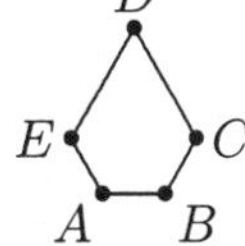

(A) 10 **(B)** $7\sqrt{3}$ **(C)** 15

(D) $9\sqrt{3}$ **(E)** $12\sqrt{5}$

15. For how many values of n will an n-sided regular polygon have interior angles with integral degree measures?

(A) 16 **(B)** 18 **(C)** 20 **(D)** 22 **(E)** 24

16. Consider the non-decreasing sequence of positive integers

$$1, 2, 2, 3, 3, 3, 4, 4, 4, 4, 5, 5, 5, 5, 5, \ldots$$

in which the nth positive integer appears n times. The remainder when the 1993rd term is divided by 5 is

(A) 0 **(B)** 1 **(C)** 2 **(D)** 3 **(E)** 4

17. Amy painted a dart board over a square clock face using the "hour positions" as boundaries. [See figure.] If t is the area of one of the eight triangular regions such as that between 12 o'clock and 1 o'clock, and q is the area of one of the four corner quadrilaterals such as that between 1 o'clock and 2 o'clock, then $\frac{q}{t} =$

(A) $2\sqrt{3} - 2$ **(B)** $\frac{3}{2}$ **(C)** $\frac{\sqrt{5}+1}{2}$ **(D)** $\sqrt{3}$ **(E)** 2

18. Al and Barb start their new jobs on the same day. Al's schedule is 3 work-days followed by 1 rest-day. Barb's schedule is 7 work-days followed by 3 rest-days. On how many of their first 1000 days do both have rest-days on the same day?

(A) 48 **(B)** 50 **(C)** 72 **(D)** 75 **(E)** 100

19. How many ordered pairs (m, n) of positive integers are solutions to

$$\frac{4}{m} + \frac{2}{n} = 1?$$

(A) 1 **(B)** 2 **(C)** 3 **(D)** 4 **(E)** more than 4

20. Consider the equation $10z^2 - 3iz - k = 0$, where z is a complex variable and $i^2 = -1$. Which of the following statements is true?

(A) For all positive real numbers k, both roots are pure imaginary.

(B) For all negative real numbers k, both roots are pure imaginary.

(C) For all pure imaginary numbers k, both roots are real and rational.

(D) For all pure imaginary numbers k, both roots are real and irrational.

(E) For all complex numbers k, neither root is real.

21. Let $a_1, a_2, \ldots, a_k$ be a finite arithmetic sequence with

$$a_4 + a_7 + a_{10} = 17 \qquad \text{and}$$

$$a_4 + a_5 + a_6 + \cdots + a_{12} + a_{13} + a_{14} = 77.$$

If $a_k = 13$, then $k =$

(A) 16 **(B)** 18 **(C)** 20 **(D)** 22 **(E)** 24

22. Twenty cubical blocks are arranged as shown. First, 10 are arranged in a triangular pattern; then a layer of 6, arranged in a triangular pattern, is centered on the 10; then a layer of 3, arranged in a triangular pattern, is centered on the 6; and finally one block is centered on top of the third layer. The blocks in the bottom layer are numbered 1 through 10 in some order. Each block in layers 2, 3 and 4 is assigned the number which is the sum of the numbers assigned to the three blocks on which it rests. Find the smallest possible number which could be assigned to the top block.

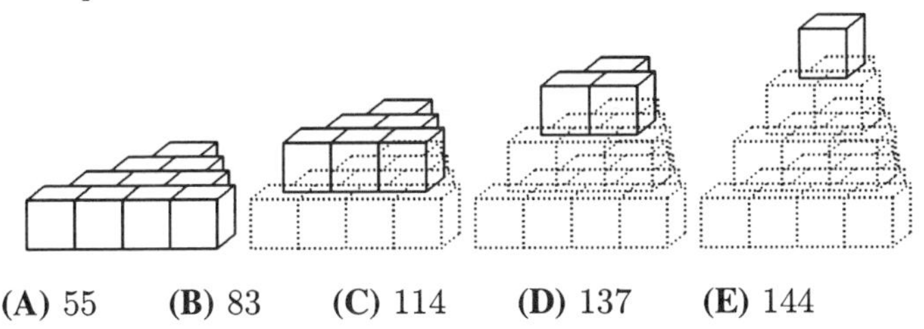

(A) 55 **(B)** 83 **(C)** 114 **(D)** 137 **(E)** 144

23. Points A, B, C and D are on a circle of diameter 1, and X is on diameter $\overline{AD}$. If $BX = CX$ and $3\angle BAC = \angle BXC = 36°$, then $AX =$

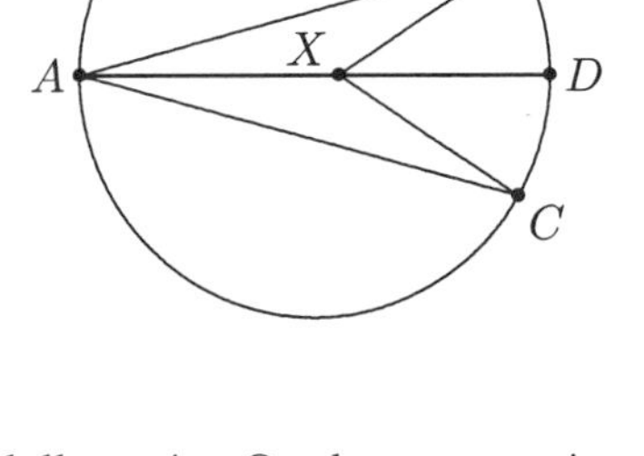

(A) $\cos 6° \cos 12° \sec 18°$

(B) $\cos 6° \sin 12° \csc 18°$

(C) $\cos 6° \sin 12° \sec 18°$

(D) $\sin 6° \sin 12° \csc 18°$

(E) $\sin 6° \sin 12° \sec 18°$

24. A box contains 3 shiny pennies and 4 dull pennies. One by one, pennies are drawn at random from the box and not replaced. If the probability is a/b that it will take more than four draws until the third shiny penny appears and a/b is in lowest terms, then $a + b =$

(A) 11 **(B)** 20 **(C)** 35 **(D)** 58 **(E)** 66

25. Let S be the set of points on the rays forming the sides of a $120°$ angle, and let P be a fixed point inside the angle on the angle bisector. Consider all distinct equilateral triangles PQR with Q and R in S.

(Points Q and R may be on the same ray, and switching the names of Q and R does not create a distinct triangle.) There are

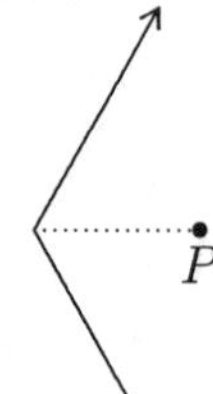

(A) exactly 2 such triangles.

(B) exactly 3 such triangles.

(C) exactly 7 such triangles.

(D) exactly 15 such triangles.

(E) more than 15 such triangles.

26. Find the largest positive value attained by the function

$$f(x) = \sqrt{8x - x^2} - \sqrt{14x - x^2 - 48}, \quad x \text{a real number.}$$

(A) $\sqrt{7} - 1$ **(B)** 3 **(C)** $2\sqrt{3}$ **(D)** 4 **(E)** $\sqrt{55} - \sqrt{5}$

27. The sides of $\triangle ABC$ have lengths 6, 8 and 10. A circle with center P and radius 1 rolls around the inside of $\triangle ABC$, always remaining tangent to at least one side of the triangle. When P first returns to its original position, through what distance has P traveled?

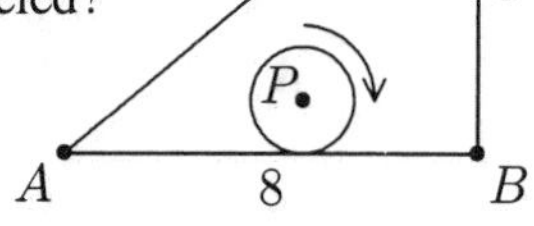

(A) 10 **(B)** 12 **(C)** 14

(D) 15 **(E)** 17

28. How many triangles with positive area are there whose vertices are points in the xy-plane whose coordinates are integers (x, y) satisfying $1 \leq x \leq 4$ and $1 \leq y \leq 4$?

(A) 496 **(B)** 500 **(C)** 512 **(D)** 516 **(E)** 560

29. Which of the following sets could NOT be the lengths of the external diagonals of a right rectangular prism [a "box"]? (An *external diagonal* is a diagonal of one of the rectangular faces of the box.)

(A) $\{4, 5, 6\}$ **(B)** $\{4, 5, 7\}$ **(C)** $\{4, 6, 7\}$

(D) $\{5, 6, 7\}$ **(E)** $\{5, 7, 8\}$

30. Given $0 \le x_0 < 1$, let

$$x_n = \begin{cases} 2x_{n-1} & \text{if } 2x_{n-1} < 1 \\ 2x_{n-1} - 1 & \text{if } 2x_{n-1} \ge 1 \end{cases}$$

for all integers $n > 0$. For how many x_0 is it true that $x_0 = x_5$?

(A) 0 **(B)** 1 **(C)** 5 **(D)** 31 **(E)** infinitely many

Some Comments on the Distractors

The distribution of responses by the top-scoring 0.94% participants in the 44th AHSME is on the next page. The data on the distractors that simply jumps out of the page are (A) and (B) on problem 8 and (B) on problem 25.

- On problem 8, over half of these students chose one of either (A) or (B). There are three pairs of circles tangent to the given circles. Those who chose (A) saw only one of these pairs, and those who chose (B) saw two of the pairs but not all three.
- On problem 25, almost three times as many students chose distractor (B) in preference to the correct answer. Those who chose (B) probably saw the two equilateral triangles with Q or R at the vertex O of the $120°$ angle, together with the equilateral triangle which has $OQ = OR$ and Q and R on opposite sides of the given angle.

Answers and Response Distribution

For the 44th AHSME, 347,720 copies of the examination were distributed to students at 5694 schools. There were 2 perfect papers. For the 3273 students named to the national Honor Roll for this examination, the following table lists the percent who gave the correct answer to each question. The percentages of responses to the other answers is also given.

	ANSWER				
#1	(D) 98.04	(A) 0.09	(B) 0.09	(C) 0.43	(E) 1.41
#2	(D) 98.41	(A) 0.06	(B) 0.06	(C) 0.40	(E) 0.00
#3	(E) 98.99	(A) 0.21	(B) 0.00	(C) 0.06	(D) 0.27
#4	(E) 98.01	(A) 0.43	(B) 0.58	(C) 0.06	(D) 0.03
#5	(A) 99.36	(B) 0.09	(C) 0.09	(D) 0.21	(E) 0.09
#6	(B) 87.44	(A) 0.31	(C) 0.52	(D) 0.18	(E) 1.41
#7	(E) 93.77	(A) 0.40	(B) 0.34	(C) 0.52	(D) 0.00
#8	(D) 33.00	(A) 28.90	(B) 28.90	(C) 0.37	(E) 0.40
#9	(D) 91.75	(A) 0.27	(B) 0.27	(C) 2.99	(E) 1.80
#10	(C) 88.94	(A) 0.18	(B) 3.30	(D) 0.21	(E) 1.89
#11	(A) 89.67	(B) 0.27	(C) 0.21	(D) 0.06	(E) 0.24
#12	(E) 91.45	(A) 0.95	(B) 0.55	(C) 0.12	(D) 1.71
#13	(D) 58.94	(A) 4.49	(B) 4.49	(C) 0.18	(E) 0.70
#14	(B) 89.58	(A) 0.06	(C) 0.00	(D) 0.55	(E) 0.09
#15	(D) 38.07	(A) 4.40	(B) 4.40	(C) 8.49	(E) 7.24
#16	(D) 55.64	(A) 1.16	(B) 1.16	(C) 4.16	(E) 4.67
#17	(A) 57.90	(B) 0.43	(C) 0.49	(D) 0.79	(E) 8.86
#18	(E) 83.26	(A) 0.12	(B) 1.44	(C) 0.15	(D) 1.01
#19	(D) 58.42	(A) 1.65	(B) 1.65	(C) 6.81	(E) 5.47
#20	(B) 25.97	(A) 3.30	(C) 0.12	(D) 0.24	(E) 3.33
#21	(B) 41.06	(A) 0.95	(C) 0.55	(D) 0.15	(E) 0.43
#22	(C) 48.40	(A) 0.18	(B) 1.53	(D) 0.27	(E) 0.58
#23	(B) 6.72	(A) 0.27	(C) 1.07	(D) 0.52	(E) 0.09
#24	(E) 19.89	(A) 0.46	(B) 0.12	(C) 0.58	(D) 0.58
#25	(E) 6.81	(A) 5.04	(B) 18.82	(C) 0.46	(D) 0.06
#26	(C) 16.96	(A) 2.14	(B) 0.89	(D) 0.89	(E) 1.07
#27	(B) 14.91	(A) 0.58	(C) 1.62	(D) 2.08	(E) 1.50
#28	(D) 6.20	(A) 0.18	(B) 0.18	(C) 0.95	(E) 4.46
#29	(B) 12.31	(A) 0.46	(C) 0.34	(D) 0.12	(E) 0.27
#30	(D) 3.64	(A) 2.51	(B) 2.51	(C) 0.95	(E) 0.89

45 AHSME

1. $4^4 \cdot 9^4 \cdot 4^9 \cdot 9^9 =$

(A) 13^{13} **(B)** 13^{36} **(C)** 36^{13} **(D)** 36^{36} **(E)** 1296^{26}

2. A large rectangle is partitioned into four rectangles by two segments parallel to its sides. The areas of three of the resulting rectangles are shown. What is the area of the fourth rectangle?

6	14
?	35

(A) 10 **(B)** 15 **(C)** 20 **(D)** 21 **(E)** 25

3. How many of the following are equal to $x^x + x^x$ for all $x > 0$?

I: $2x^x$ **II**: x^{2x} **III**: $(2x)^x$ **IV**: $(2x)^{2x}$

(A) 0 **(B)** 1 **(C)** 2 **(D)** 3 **(E)** 4

4. In the xy-plane, the segment with endpoints $(-5, 0)$ and $(25, 0)$ is the diameter of a circle. If the point $(x, 15)$ is on the circle, then $x =$

(A) 10 **(B)** 12.5 **(C)** 15 **(D)** 17.5 **(E)** 20

5. Pat intended to multiply a number by 6 but instead divided by 6. Pat then meant to add 14 but instead subtracted 14. After these mistakes,

the result was 16. If the correct operations had been used, the value produced would have been

(A) less than 400 **(B)** between 400 and 600
(C) between 600 and 800 **(D)** between 800 and 1000
(E) greater than 1000

6. In the sequence

$$\ldots, a, b, c, d, 0, 1, 1, 2, 3, 5, 8, \ldots$$

each term is the sum of the two terms to its left. Find a.

(A) -3 **(B)** -1 **(C)** 0 **(D)** 1 **(E)** 3

7. Squares $ABCD$ and $EFGH$ are congruent, $AB = 10$, and G is the center of square $ABCD$. The area of the region in the plane covered by these squares is

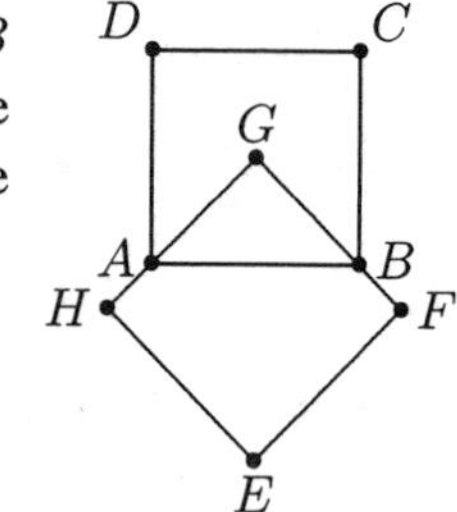

(A) 75 **(B)** 100 **(C)** 125
(D) 150 **(E)** 175

8. In the polygon shown, each side is perpendicular to its adjacent sides, and all 28 of the sides are congruent. The perimeter of the polygon is 56. The area of the region bounded by the polygon is

(A) 84 **(B)** 96 **(C)** 100 **(D)** 112 **(E)** 196

9. If $\angle A$ is four times $\angle B$, and the complement of $\angle B$ is four times the complement of $\angle A$, then $\angle B =$

(A) $10°$ **(B)** $12°$ **(C)** $15°$ **(D)** $18°$ **(E)** $22.5°$

10. For distinct real numbers x and y, let $M(x, y)$ be the larger of x and y and let $m(x, y)$ be the smaller of x and y. If

$$a < b < c < d < e$$

then

$$M(M(a, m(b, c)), m(d, m(a, e))) =$$

(A) a **(B)** b **(C)** c **(D)** d **(E)** e

11. Three cubes of volume 1, 8 and 27 are glued together at their faces. The smallest possible surface area of the resulting configuration is

(A) 36 **(B)** 56 **(C)** 70 **(D)** 72 **(E)** 74

12. If $i^2 = -1$, then $\left(i - i^{-1}\right)^{-1} =$

(A) 0 **(B)** $-2i$ **(C)** $2i$ **(D)** $-i/2$ **(E)** $i/2$

13. In triangle ABC, $AB = AC$. If there is a point P strictly between A and B such that $AP = PC = CB$, then $\angle A =$

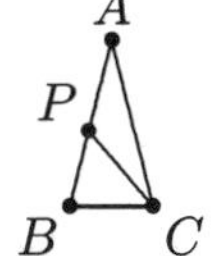

(A) 30° **(B)** 36° **(C)** 48° **(D)** 60° **(E)** 72°

14. Find the sum of the arithmetic series

$$20 + 20\frac{1}{5} + 20\frac{2}{5} + \cdots + 40.$$

(A) 3000 **(B)** 3030 **(C)** 3150 **(D)** 4100 **(E)** 6000

15. For how many n in $\{1, 2, 3, \ldots, 100\}$ is the tens digit of n^2 odd?

(A) 10 **(B)** 20 **(C)** 30 **(D)** 40 **(E)** 50

16. Some marbles in a bag are red and the rest are blue. If one red marble is removed, then one-seventh of the remaining marbles are red. If two blue marbles are removed instead of one red, then one-fifth of the remaining marbles are red. How many marbles were in the bag originally?

(A) 8 **(B)** 22 **(C)** 36 **(D)** 57 **(E)** 71

17. An 8 by $2\sqrt{2}$ rectangle has the same center as a circle of radius 2. The area of the region common to both the rectangle and the circle is

(A) 2π **(B)** $2\pi + 2$ **(C)** $4\pi - 4$ **(D)** $2\pi + 4$

(E) $4\pi - 2$

18. Triangle ABC is inscribed in a circle, and $\angle B = \angle C = 4\angle A$. If B and C are adjacent vertices of a regular polygon of n sides inscribed in this circle, then $n =$

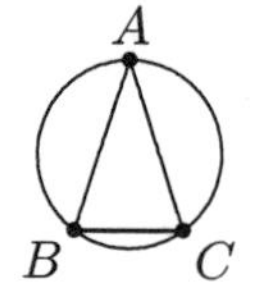

(A) 5 (B) 7 (C) 9 (D) 15 (E) 18

19. Label one disk "1", two disks "2", three disks "3", ..., fifty disks "50". Put these $1+2+3+\cdots+50 = 1275$ labeled disks in a box. Disks are then drawn from the box at random without replacement. The minimum number of disks that must be drawn to guarantee drawing at least ten disks with the same label is

(A) 10 (B) 51 (C) 415 (D) 451 (E) 501

20. Suppose x, y, z is a geometric sequence with common ratio r and $x \neq y$. If $x, 2y, 3z$ is an arithmetic sequence, then r is

(A) $\frac{1}{4}$ (B) $\frac{1}{3}$ (C) $\frac{1}{2}$ (D) 2 (E) 4

21. Find the number of counterexamples to the statement:

"If N is an odd positive integer the sum of whose digits is 4 and none of whose digits is 0, then N is prime."

(A) 0 (B) 1 (C) 2 (D) 3 (E) 4

22. Nine chairs in a row are to be occupied by six students and Professors Alpha, Beta and Gamma. These three professors arrive before the six students and decide to choose their chairs so that each professor will be between two students. In how many ways can Professors Alpha, Beta and Gamma choose their chairs?

(A) 12 (B) 36 (C) 60 (D) 84 (E) 630

23. In the xy-plane, consider the L-shaped region bounded by horizontal and vertical segments with vertices at $(0,0)$, $(0,3)$, $(3,3)$, $(3,1)$, $(5,1)$ and $(5,0)$. The slope of the line through the origin that divides the area of this region exactly in half is

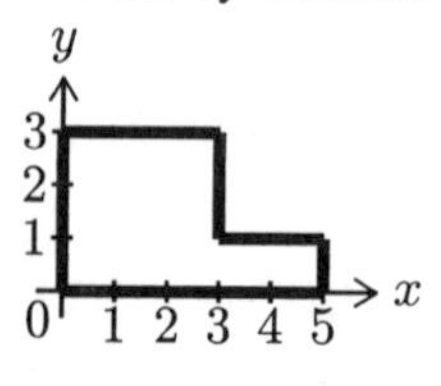

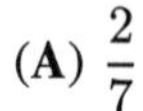
(A) $\frac{2}{7}$ (B) $\frac{1}{3}$ (C) $\frac{2}{3}$ (D) $\frac{3}{4}$ (E) $\frac{7}{9}$

24. A sample consisting of five observations has an arithmetic mean of 10 and a median of 12. The smallest value that the range (largest observation minus smallest) can assume for such a sample is

(A) 2 **(B)** 3 **(C)** 5 **(D)** 7 **(E)** 10

25. If x and y are non-zero real numbers such that

$$|x| + y = 3 \qquad \text{and} \qquad |x|\, y + x^3 = 0,$$

then the integer nearest to $x - y$ is

(A) -3 **(B)** -1 **(C)** 2 **(D)** 3 **(E)** 5

26. A regular polygon of m sides is exactly enclosed (no overlaps, no gaps) by m regular polygons of n sides each. (Shown here for $m = 4$, $n = 8$.) If $m = 10$, what is the value of n?

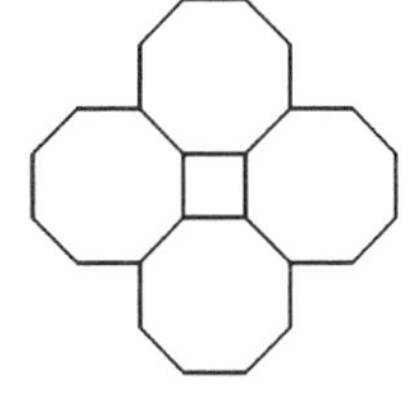

(A) 5 **(B)** 6 **(C)** 14 **(D)** 20 **(E)** 26

27. A bag of popping corn contains $2/3$ white kernels and $1/3$ yellow kernels. Only $1/2$ of the white kernels will pop, whereas $2/3$ of the yellow ones will pop. A kernel is selected at random from the bag, and pops when placed in the popper. What is the probability that the kernel selected was white?

(A) $\frac{1}{2}$ **(B)** $\frac{5}{9}$ **(C)** $\frac{4}{7}$ **(D)** $\frac{3}{5}$ **(E)** $\frac{2}{3}$

28. In the xy-plane, how many lines whose x-intercept is a positive prime number and whose y-intercept is a positive integer pass through the point (4,3)?

(A) 0 **(B)** 1 **(C)** 2 **(D)** 3 **(E)** 4

29. Points A, B and C on a circle of radius r are situated so that $AB = AC$, $AB > r$, and the length of minor arc BC is r. If angles are measured in radians, then $AB/BC =$

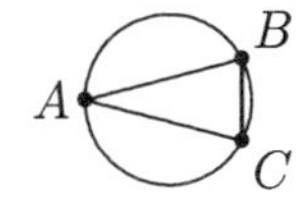

(A) $\frac{1}{2}\csc\frac{1}{4}$ **(B)** $2\cos\frac{1}{2}$ **(C)** $4\sin\frac{1}{2}$ **(D)** $\csc\frac{1}{2}$ **(E)** $2\sec\frac{1}{2}$

30. When n standard 6-sided dice are rolled, the probability of obtaining a sum of 1994 is greater than zero and is the same as the probability of obtaining a sum of S. The smallest possible value of S is

    (A) 333 **(B)** 335 **(C)** 337 **(D)** 339 **(E)** 341

Some Comments on the Distractors

The data on responses for honor roll students on the 45th AHSME, listed on the next page, might not be directly comparable to the similar tables following the five AHSMEs earlier in this book. The percentage of Honor Roll students on the 45th AHSME was over six times the average of that percent for the 40th through 44th AHSME. On this easier examination and for this larger number of Honor Roll students, no distractor was as popular as the correct choice. However, on problems 11, 21, and 30, over 10% of the honor roll students chose some particular distractor.

- On problem 11, students probably chose (E), which is 2 larger than the correct response, because they did not see that the smallest cube could be stuck simultaneously to both of the larger cubes.
- On problem 21, those who chose (B) missed the compositeness of one of the numbers, and 1111 is quite possibly the integer they thought was prime.

There was additional data on distractors available for the first time on the 45th AHSME. Be sure to look at the table and comments on the pages following the tabulation of honor roll students' answers.

Answers and Response Distribution

For the 45th AHSME, 339,890 copies of the examination were distributed to students at 5372 schools. There were 99 perfect papers. For the 20,492 students named to the national Honor Roll for this examination, the following table lists the percent who gave the correct answer to each question. The percentages of responses to the other answers is also given.

	ANSWER				
#1	: (C) 97.28	(A) 0.39	(B) 0.32	(D) 0.50	(E)0.71
#2	: (B) 98.23	(A) 0.42	(C) 0.03	(D) 0.40	(E)0.04
#3	: (B) 95.44	(A) 1.01	(C) 1.59	(D) 0.05	(E)0.03
#4	: (A) 94.05	(B) 0.48	(C) 2.17	(D) 0.15	(E)0.32
#5	: (E) 97.20	(A) 1.58	(B) 0.16	(C) 0.48	(D)0.29
#6	: (A) 95.39	(B) 0.41	(C) 1.97	(D) 0.14	(E)1.03
#7	: (E) 94.26	(A) 0.33	(B) 0.12	(C) 0.21	(D)3.88
#8	: (C) 98.27	(A) 0.14	(B) 0.82	(D) 0.19	(E)0.18
#9	: (D) 95.83	(A) 0.29	(B) 0.29	(C) 0.25	(E)0.39
#10	: (B) 61.59	(A) 1.02	(C) 0.44	(D) 0.47	(E)0.18
#11	: (D) 67.44	(A) 1.42	(B) 1.42	(C) 2.14	(E)12.86
#12	: (D) 82.10	(A) 1.46	(B) 1.46	(C) 0.46	(E)0.97
#13	: (B) 63.45	(A) 5.07	(C) 0.60	(D) 2.78	(E)1.35
#14	: (B) 77.08	(A) 3.19	(C) 2.06	(D) 0.42	(E)1.05
#15	: (B) 73.42	(A) 1.40	(C) 0.53	(D) 0.35	(E)4.83
#16	: (B) 78.33	(A) 0.74	(C) 0.61	(D) 1.57	(E)0.10
#17	: (D) 29.90	(A) 1.98	(B) 1.98	(C) 3.07	(E)5.41
#18	: (C) 34.19	(A) 4.68	(B) 1.27	(D) 0.28	(E)7.47
#19	: (C) 53.01	(A) 0.90	(B) 0.47	(D) 2.76	(E)0.74
#20	: (B) 40.86	(A) 0.20	(C) 0.39	(D) 0.43	(E)0.11
#21	: (C) 33.22	(A) 7.23	(B) 12.10	(D) 3.92	(E)1.88
#22	: (C) 24.46	(A) 3.22	(B) 6.13	(D) 3.01	(E)1.51
#23	: (E) 47.82	(A) 0.55	(B) 0.35	(C) 3.66	(D)3.77
#24	: (C) 35.80	(A) 1.05	(B) 1.54	(D) 7.61	(E)1.91
#25	: (A) 24.33	(B) 1.59	(C) 0.59	(D) 1.04	(E)0.23
#26	: (A) 30.95	(B) 0.75	(C) 0.41	(D) 5.26	(E)0.36
#27	: (D) 34.48	(A) 1.35	(B) 1.35	(C) 0.20	(E)6.77
#28	: (C) 19.95	(A) 2.90	(B) 6.03	(D) 1.35	(E)1.05
#29	: (A) 5.48	(B) 0.48	(C) 0.59	(D) 0.51	(E)0.37
#30	: (C) 13.05	(A) 10.62	(B) 2.25	(D) 0.34	(E)0.11

The results for all the 1994 students are given in the table below. It might be interesting to compare this table of responses from the average student who took the AHSME with the previous table of responses from the honor roll students.

ANSWER				
#1 : (C) 57.40	(A) 1.21	(B) 0.96	(D) 2.82	(E) 31.65
#2 : (B) 83.05	(A) 1.98	(C) 1.33	(D) 2.56	(E) 0.64
#3 : (B) 59.35	(A) 5.31	(C) 12.48	(D) 2.45	(E) 2.01
#4 : (A) 50.89	(B) 3.28	(C) 11.73	(D) 1.64	(E) 3.34
#5 : (E) 76.44	(A) 13.26	(B) 1.33	(C) 3.70	(D) 0.74
#6 : (A) 72.30	(B) 3.41	(C) 8.62	(D) 1.98	(E) 4.18
#7 : (E) 59.56	(A) 3.96	(B) 3.79	(C) 2.78	(D) 8.86
#8 : (C) 63.79	(A) 2.21	(B) 3.39	(D) 6.03	(E) 5.08
#9 : (D) 59.55	(A) 2.80	(B) 2.80	(C) 1.80	(E) 4.86
#10 : (B) 13.15	(A) 2.75	(C) 1.95	(D) 1.45	(E) 2.09
#11 : (D) 18.97	(A) 4.56	(B) 4.56	(C) 5.78	(E) 15.09
#12 : (D) 29.30	(A) 6.16	(B) 6.16	(C) 2.65	(E) 2.98
#13 : (B) 25.58	(A) 15.20	(C) 2.64	(D) 11.25	(E) 1.60
#14 : (B) 25.12	(A) 6.32	(C) 7.72	(D) 4.07	(E) 3.10
#15 : (B) 29.25	(A) 4.15	(C) 2.78	(D) 3.54	(E) 22.64
#16 : (B) 28.59	(A) 6.75	(C) 4.40	(D) 5.44	(E) 0.89
#17 : (D) 7.09	(A) 2.69	(B) 2.69	(C) 6.57	(E) 7.45
#18 : (C) 8.08	(A) 12.12	(B) 4.02	(D) 1.85	(E) 4.37
#19 : (C) 13.78	(A) 8.62	(B) 6.10	(D) 5.32	(E) 4.02
#20 : (B) 9.07	(A) 1.71	(C) 3.31	(D) 3.05	(E) 1.57
#21 : (C) 14.24	(A) 15.31	(B) 12.37	(D) 6.30	(E) 4.08
#22 : (C) 6.55	(A) 16.92	(B) 15.19	(D) 5.89	(E) 2.24
#23 : (E) 12.38	(A) 1.40	(B) 3.90	(C) 14.20	(D) 8.64
#24 : (C) 10.02	(A) 7.75	(B) 3.03	(D) 6.41	(E) 5.12
#25 : (A) 14.00	(B) 5.15	(C) 2.49	(D) 3.05	(E) 0.95
#26 : (A) 8.18	(B) 1.91	(C) 3.10	(D) 35.40	(E) 1.11
#27 : (D) 11.46	(A) 7.17	(B) 7.17	(C) 1.90	(E) 14.20
#28 : (C) 7.59	(A) 6.63	(B) 7.47	(D) 3.43	(E) 4.09
#29 : (A) 1.38	(B) 2.39	(C) 1.88	(D) 1.23	(E) 1.11
#30 : (C) 3.24	(A) 20.42	(B) 2.24	(D) 1.36	(E) 0.91

The first year the AHSME was given, 1949, was three years before the first computer was sold commercially. The shear quantity of papers necessitated decentralized scoring. All papers were graded first by the examination manager at the school. The high-scoring papers from the school were forwarded to the state coordinator who verified the scores and forwarded the papers of students on the national honor roll to the central office where the executive director verified the scoring of the honor roll papers. Since these honor roll papers were the only ones arriving at the national office, it was these papers that were analyzed for the distribution of responses to the question.

The change from decentralized hand-scoring to centralized computer scoring began with the 1992 examination. All enhancements could not be implemented at the start. The first year in which the computer data on the responses of all students (not just the honor roll students) was analyzed was in 1994. That is the analysis printed above.

40 AHSME Solutions

1. **(C)** $(-1)^{5^2} + 1^{2^5} = (-1)^{25} + 1^{32} = -1 + 1 = 0.$

Comment. Since $1^n = 1$ for all n and

$$(-1)^n = \begin{cases} 1 & \text{if } n \text{ is even} \\ -1 & \text{if } n \text{ is odd,} \end{cases}$$

this problem tests knowledge of the notational convention that

$$(-1)^{5^2} \begin{cases} = (-1)^{(5^2)} & \text{and} \\ \neq \left((-1)^5\right)^2. \end{cases}$$

2. **(D)** $\sqrt{\dfrac{1}{9} + \dfrac{1}{16}} = \sqrt{\dfrac{16+9}{144}} = \sqrt{\dfrac{25}{144}} = \dfrac{5}{12}.$

Comment. The square root is **not** a linear operator; that is,

$$\sqrt{\frac{1}{9} + \frac{1}{16}} \neq \frac{1}{3} + \frac{1}{4},$$

which is the reason $7/12$ is not the correct answer.

3. **(D)** Let x be the length of the shorter side of one of the rectangles. The perimeter of each of the rectangles is $8x$, so $8x = 24$ and $x = 3$. Since each side of the square is $3x$, the area of the square is $(3x)^2 = 81$.

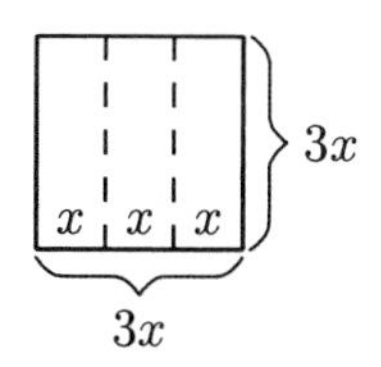

4. **(D)** Drop perpendiculars $\overline{AG}$ and $\overline{BH}$ to $\overline{DF}$. Then $GH = 4$, so

$$DG = HC = \frac{1}{2}(DC - GH) = 3.$$

Since $\overline{BH} \parallel \overline{EF}$ and B is the midpoint of DE, it follows that H is the midpoint of DF. Thus,

$$DH = DG + GH = 3 + 4$$

and $DF = 2DH = 14$, so $CF = DF - DC = 14 - 10 = 4$.

5. **(E)** For an $m \times n$ grid there are $m+1$ columns of vertical toothpicks, each n toothpicks long, so there are $(m+1)n$ vertical toothpicks. Likewise, there are $(n+1)m$ horizontal toothpicks. The total is $(m+1)n + (n+1)m$ toothpicks. In our case with $m = 10$ and $n = 20$, the number of toothpicks is $11 \cdot 20 + 21 \cdot 10 = 430$.

OR

Each of the 20×10 unit squares has four sides. Thus $4 \cdot (20 \cdot 10)$ counts each toothpick twice, except for the $2 \cdot (20 + 10)$ toothpicks on the perimeter of the rectangle, which are only counted once. Hence, the number of toothpicks is $\frac{1}{2}(4 \cdot 20 \cdot 10 + 2(20 + 10)) = 430$.

6. **(A)** The x-intercept of the line is $6/a$, and the y-intercept is $6/b$. Thus the area of the triangle is

$$\frac{1}{2} \cdot \frac{6}{a} \cdot \frac{6}{b} = \frac{18}{ab}.$$

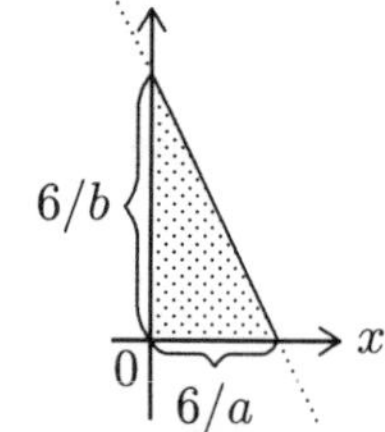

We require the area to be 6, so $\dfrac{18}{ab} = 6$, from which it follows that $ab = 3$.

OR

One might infer from the statement of the problem that the product ab is independent of the choice of lines.† As stated, it can be solved

† In early stages of the development of this examination, one of the answer choices was *Not Uniquely Determined* which would have precluded this inference, and perhaps made the problem more difficult.

by testing a couple right triangles with area 6, such as those with legs 3 and 4, or 12 and 1:

$$\begin{array}{rcc} \text{Through}: & (3,0),(0,4) & (12,0),(0,1) \\ \text{Slope}: & m=-4/3 & m=-1/12 \\ \text{Equation}: & y=-(4/3)x+4 & y=-(1/12)x+1 \\ & (4/3)x+y=4 & (1/12)x+y=1 \\ & 2x+(3/2)y=6 & (1/2)x+6y=6 \\ \Rightarrow & ab=2\cdot(3/2)=\mathbf{3} & ab=(1/2)\cdot 6=\mathbf{3} \end{array}$$

7. **(C)** Concentrate on $\triangle AHC$. Since $\overline{BM}$ is a median and $\overline{AH}$ is an altitude of $\triangle ABC$, it follows that M is the midpoint of $\overline{AC}$ and $AH \perp HC$. Thus, $\triangle AHC$ is a 30°-60°-90° triangle. Hence, $HA = \frac{1}{2}AC = AM$. But $\angle HAM = 60°$, so $\triangle AHM$ is equilateral. Thus, $\angle AHM = 60°$, and $\angle MHC = 90° - 60° = 30°$.

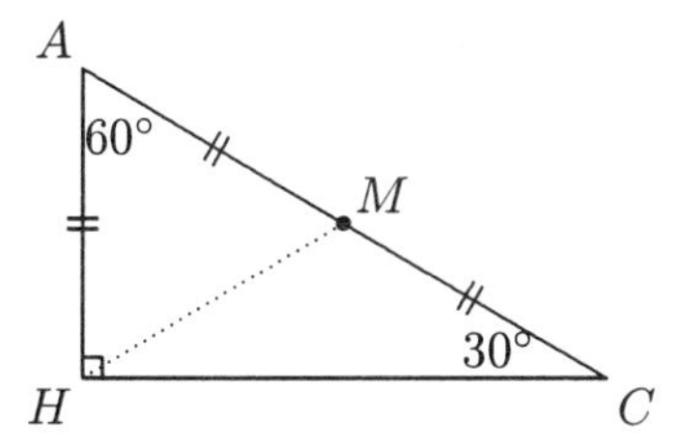

OR

As soon as $\triangle AHM$ is seen to be equilateral, we have $HM = MC$. Thus $\triangle HMC$ is isosceles, from which it follows that $\angle MHC = \angle MCH = 30°$.

OR

Since M is the midpoint of hypotenuse $\overline{AC}$ of right triangle AHC, $\overline{MH}$ and $\overline{MC}$ are radii of the circle circumscribing $\triangle AHC$. Therefore, in $\triangle HMC$ we have $MH = MC$, so $\angle MHC = \angle C = 30°$.

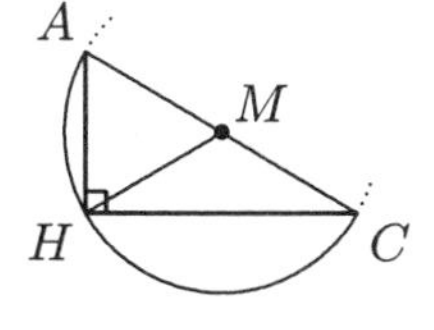

8. **(D)** Since $-n < 0$,

$$x^2 + x - n = (x-a)(x+b)$$

where a and b are positive integers. Since the coefficient of x is 1, $b = a + 1$. Complete this table:

a:	1	2	...	9	10
b:	2	3	...	10	11
$n{=}ab$:	$1{\cdot}2{=}2$	$2{\cdot}3{=}6$	...	$9{\cdot}10{=}90$	$10{\cdot}11{=}110$

There are 9 values for n between 1 and 100.

OR

By the quadratic formula, $x^2+x-n=(x-x_1)(x-x_2)$ if and only if

$$x_1, x_2 = \frac{-1 \pm \sqrt{1+4n}}{2}.$$

Since x_1 and x_2 are integers, $1+4n=m^2$ for some positive integer m. Since $1+4n$ is odd, m must be odd; i.e., $m=2k+1$ for some integer k. Every odd square is of the form $1+4n$ because

$$m^2=(2k+1)^2=4k^2+4k+1=4(k^2+k)+1.$$

Since $1 \le n \le 100$, the answer is the number of odd squares between $5=1+4{\cdot}1$ and $401=1+4{\cdot}100$. There are nine odd squares in this range, 3^2, 5^2, 7^2, 9^2, 11^2, 13^2, 15^2, 17^2 and 19^2.

9. **(B)** The possible monograms are

$$ABZ, ACZ, ADZ, \ldots, WXZ, WYZ, XYZ.$$

Any two-element subset of the first 25 letters of the alphabet, when used in alphabetical order, will produce a suitable monogram when combined with Z. For example $\{L, J\}=\{J, L\}$ will produce JLZ. Furthermore, to every suitable monogram there corresponds exactly one two-element subset of $\{A, B, C, \ldots, Y\}$. Thus, the answer is the number of two-element subsets that can be formed from a set of 25 letters, and there are $\binom{25}{2}=300$ such subsets.

OR

The last initial is fixed at Z. If the first initial is A, the second initial must be one of $B, C, D, \ldots, Y$, so there are 24 choices for the second. If the first initial is B, there are 23 choices for the second initial: of $C, D, E, \ldots, Y$. Continuing in this way we see that the number of monograms is

$$24+23+22+\cdots+1.$$

Use the formula

$$1+2+\cdots+n=\frac{n(n+1)}{2}$$

to obtain the answer $(24\cdot 25)/2=300$.

10. **(C)** Compute the sequence $u_1, u_2, u_3, \ldots$ to look for a pattern:

$$u_1 = a$$
$$u_2 = \frac{-1}{a+1}$$
$$u_3 = \frac{-1}{\frac{-1}{a+1}+1} = \frac{-(a+1)}{a}$$
$$u_4 = \frac{-1}{-\frac{(a+1)}{a}+1} = \frac{-a}{-1} = a$$

Hence $a = u_1 = u_4 = u_7 = \cdots = u_{16} = \cdots$. Among the given answer choices, only the subscript 16 appears.

We can verify that none of the other given choices for n yield $u_n = a$ for all a by noting that when $a = 1$:

$$u_2 = \frac{-1}{2} \quad \text{so} \quad u_{14} = u_{17} = -\frac{1}{2} \neq 1 = a$$
$$u_3 = \frac{-1}{1/2} \quad \text{so} \quad u_{15} = u_{18} = -2 \neq 1 = a.$$

11. **(A)** To make a as large as possible, b must be chosen as large as possible. Similarly, b and c will attain their largest possible values only when c and d, respectively, are as large as possible.
The largest possible value for d is 99.
Since $c < 4{\cdot}99 = 396$, the largest value for c is 395.
Since $b < 3{\cdot}395 = 1185$, the largest value for b is 1184.
Since $a < 2{\cdot}1184 = 2368$, the largest value for a is 2367.

OR

In general, from the first three inequalities, the maximum values of c, b and a in terms of d must be:

$$c = 4d - 1$$
$$b = 3c - 1 = 3(4d-1) - 1 = 12d - 4$$
$$a = 2b - 1 = 2(12d-4) - 1 = 24d - 9$$

Since $d = 100 - 1$ is the largest possible value for d, it follows that the largest possible value for a is

$$a = 24(100-1) - 9 = 2400 - 33 = 2367.$$

12. **(C)** Since the vehicles moving in each direction are traveling at 60 miles per hour, in five minutes all the westbound vehicles have moved

five miles farther west, and the eastbound driver has moved five miles farther east. Thus, in five minutes the eastbound driver will actually count the number of westbound vehicles in a ten-mile section of highway:

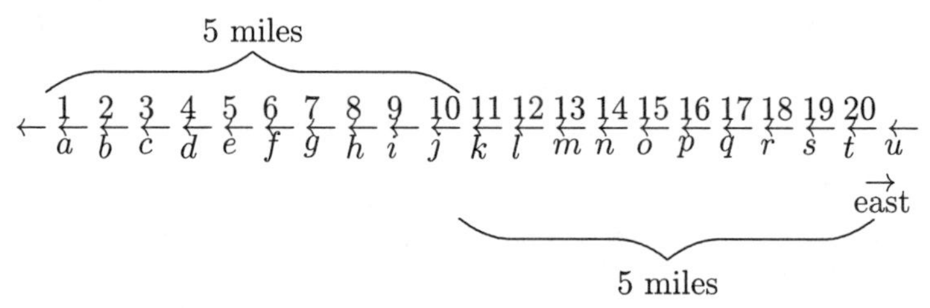

Since 20 vehicles are in a ten-mile section, there will be 200 vehicles in a 100-mile section of highway.

13. **(B)** The shaded figure in the problem is a rhombus. Each side has length $(1/\sin\alpha)$, which can be observed in the right triangle indicated in the figure. The height of the rhombus is 1, which is the width of each strip. The area of the rhombus is $base \cdot height = 1\cdot(1/\sin\alpha)$.

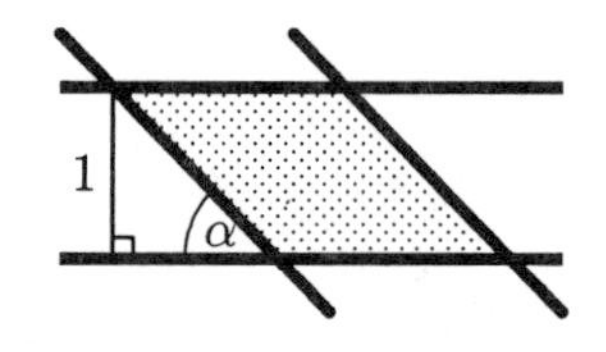

OR

The area of a parallelogram is the product of the lengths of two adjacent sides and the sine of the angle between them. Therefore, the area of this rhombus is $(1/\sin\alpha)^2\cdot\sin\alpha$.

14. **(B)** Use the definitions of the tangent and cotangent functions and the identity for the cosine of the difference of two angles to obtain

$$\begin{aligned}\cot 10+\tan 5 &= \frac{\cos 10}{\sin 10}+\frac{\sin 5}{\cos 5}\\ &= \frac{\cos 10\cos 5+\sin 10\sin 5}{\sin 10\cos 5}=\frac{\cos(10-5)}{\sin 10\cos 5}\\ &= \frac{\cos(5)}{\sin 10\cos 5}=\frac{1}{\sin 10}=\csc 10.\end{aligned}$$

Note. This is an instance of the identity

$$\cot 2x + \tan x = \csc 2x.$$

To prove this general identity, imitate that above substituting $2x$ for 10 and x for 5.

15. **(E)** Apply the *Law of Cosines* to $\triangle BAC$:

$$BC^2 = BA^2 + AC^2 - 2(BA)(AC)\cos A$$

$$49 = 25 + 81 - 2(5)(9)\cos A.$$

Thus $\cos A = 19/30$. Let H be the foot of the altitude from B. Then

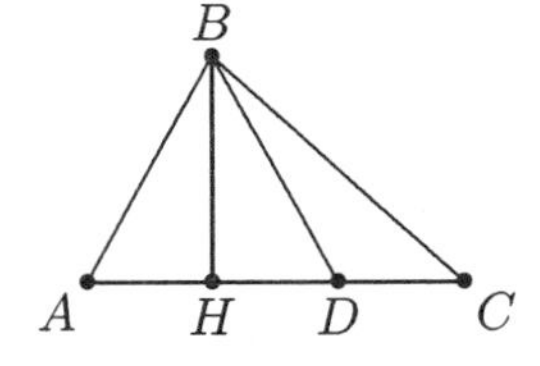

$$AD = 2 \cdot AH = 2(AB)\cos A = \frac{19}{3},$$

$$DC = AC - AD = 8/3, \text{ and}$$

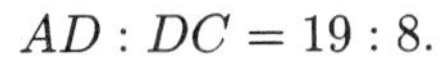

$$AD : DC = 19 : 8.$$

OR

Apply the *Pythagorean Theorem* to triangles AHB and CHB:

$$5^2 - AH^2 = BH^2 = 7^2 - (9 - AH)^2.$$

Solve to find that $AH = 19/6$ so $AD = 2 \cdot AH = 19/3$. Thus

$$AD : DC = AD : (9 - AD) = \frac{19}{3} : \frac{8}{3} = 19 : 8.$$

16. **(B)** The slope of the line segment is

$$\frac{281 - 17}{48 - 3} = \frac{264}{45} = \frac{88}{15},$$

so the equation of the line containing the segment is

$$y = 17 + \frac{88}{15}(x - 3).$$

Thus, the lattice point (x, y) will be on the line segment if and only if

- x and y are integers,
- $3 \le x \le 48$, and
- $y = 17 + 88(x - 3)/15$.

Since 88 and 15 are relatively prime, $x-3$ must be a multiple of 15; that is $x = 3, 18, 33, \ldots$. The four lattice points are, therefore,

$$(3,17),\ (18,105),\ (33,193) \text{ and } (48,281).$$

OR

The number of lattice points on the segment from the origin to (a, b) is one more than the greatest common factor of a and b. The number of lattice points on the segment $\overline{(3,17)(48,281)}$ is the same as the number on its translation to the segment $\overline{(0,0)(45,264)}$. The greatest common factor of 45 and 264 is 3, so there are 4 lattice points on the segment.

17. **(D)** Let t denote the length of each side of the triangle and s denote the length of each side of the square. Then $3t = 4s + 1989$ and $d = t - s$. Hence

$$d = \frac{4s + 1989}{3} - s = \frac{s}{3} + 663.$$

Because $s > 0$, $d \leq 663$ is impossible. However, s may take on any positive value, so all integral values of d that exceed 663 (as well as many non-integral values) are possible. Thus, only $1, 2, \ldots, 663$ are excluded as positive integer values for d.

OR

For any (s, t),

$$3t = 4s + 1989.$$

Subtract $\quad 3(663) = 4(0) + 1989,$

to obtain $\quad 3(t - 663) = 4s.$

Since 3 and 4 are relatively prime, when s and t are both integers, $t - 663$ must be a multiple of 4 and s must be a multiple of 3. In fact, all positive integer solutions to $3t = 4s + 1989$ are of the form:

$$s = 3k,\ t = 663 + 4k,\ k = 1, 2, 3, \ldots.$$

Thus, possible values for $d = t - s$ are

$$d = (663 + 4k) - 3k = 663 + k,\ k = 1, 2, 3, \ldots,$$

so d can be any integer greater than 663. Since $3k = s > 0$, no smaller value for d is possible.

18. **(B)** Rationalize the denominator of the third term in the given expression:

$$\frac{1}{x+\sqrt{x^2+1}} \cdot \frac{x-\sqrt{x^2+1}}{x-\sqrt{x^2+1}} = \frac{\sqrt{x^2+1}-x}{1}.$$

Thus $x+\sqrt{x^2+1}$ and $\sqrt{x^2+1}-x$ are reciprocals. Hence the given expression

$$\begin{aligned} x+\sqrt{x^2+1}-\frac{1}{x+\sqrt{x^2+1}} &= x+\sqrt{x^2+1}-\left(\sqrt{x^2+1}-x\right) \\ &= 2x \end{aligned}$$

which is rational if and only if x is rational.

To show that this is the only answer, show that each of the other sets is different from the set of rational numbers:

- **(A)** and **(C)** are not the set of rational numbers.
- The rational number $x=1$ is not in the set described by **(D)** or **(E)** since neither $\sqrt{2}$ nor $1+\sqrt{2}$ is rational.

19. **(E)** Let R be the radius of the circle. Then the circumference of the circle is $2\pi R = 3+4+5$ so $R = 6/\pi$. The central angle subtended by the arc of length 5 measures $5/(6/\pi) = 5\pi/6$ radians. Likewise, the angles subtended by the arcs of length 4 and 3 have measures $2\pi/3$ and $\pi/2$ radians, respectively. The area of the given triangle is the sum of the areas of the three triangles into which it is partitioned by the radii to its vertices. A formula for the area of an isosceles triangle with two sides of length R adjacent to an angle of measure θ is $(1/2)R^2 \sin\theta$, so the answer is

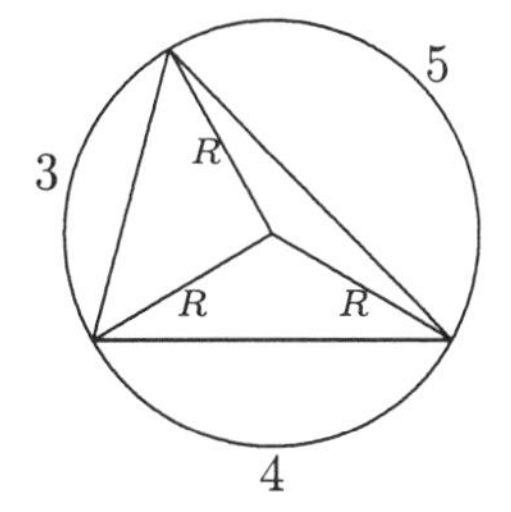

$$\begin{aligned} \frac{1}{2}R^2\left(\sin\frac{5\pi}{6}+\sin\frac{2\pi}{3}+\sin\frac{\pi}{2}\right) &= \frac{1}{2}\left(\frac{6}{\pi}\right)^2\left(\frac{1}{2}+\frac{\sqrt{3}}{2}+1\right) \\ &= \frac{1}{2}\left(\frac{36}{\pi^2}\right)\left(\frac{3+\sqrt{3}}{2}\right) \\ &= \frac{9}{\pi^2}(3+\sqrt{3}). \end{aligned}$$

Comment. When the angle θ between two sides a and b of a triangle is acute, the definition of the sine function shows that the altitude to side a is $b\sin\theta$, and hence the area of the triangle is $(1/2)ab\sin\theta$. When θ is obtuse, it is necessary to use an additional fact, that the sines of supplementary angles are equal, to obtain the same result.

20. **(B)** First convert the two given equations to inequalities:

 - *The following three statements are equivalent:*

$$\lfloor\sqrt{x}\rfloor = 12, \qquad 12 \le \sqrt{x} < 13, \qquad 144 \le x < 169$$

 - *The following four statements are equivalent:*

$$\lfloor\sqrt{100x}\rfloor = 120, \qquad 120 \le \sqrt{100x} < 121,$$

$$12 \le \sqrt{x} < 12.1, \qquad 144 \le x < 146.41$$

Thus, the probability, p, that a number selected at random in the interval $[144, 169)$ is also in the interval $[144, 146.41)$ is

$$p = \frac{146.41 - 144}{169 - 144} = \frac{2.41}{25} = \frac{241}{2500}.$$

Note. Choosing x between 144 and 169 uniformly at random is not the same as choosing $\sqrt{x}$ uniformly at random between 12 and 13. This is what makes choice **(C)** wrong.

21. **(C)** Let the total area be 100 and let each red segment on the border of the flag be of length x. Then the four white triangles can be placed together to form a white square of area $100-36 = 64$ and side $10-2x$.

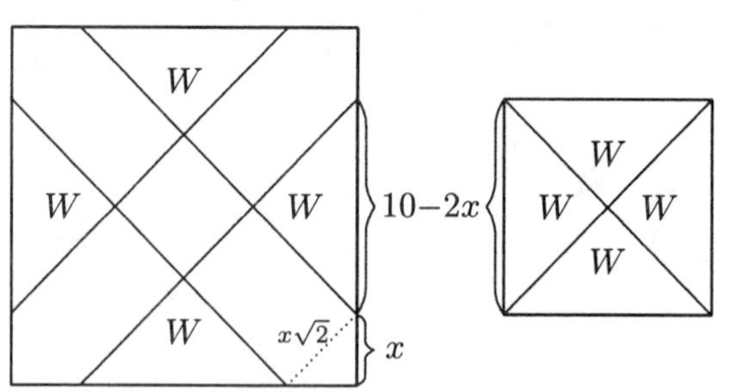

Since $(10-2x)^2 = 64$, we have $x = 1$. Thus the blue area is $(x\sqrt{2})^2 = 2$, which is 2% of the total area.

OR

First note that the flag can be cut into four congruent isosceles right triangles by the two diagonals of the flag and that the percent of red, white, and blue areas in each of these triangles is the same as that in the flag. Then form a square translating $\triangle DCE$ down and attaching it to $\triangle ABE$ along their hypotenuses $\overline{AB}$ and $\overline{DC}$ as shown.

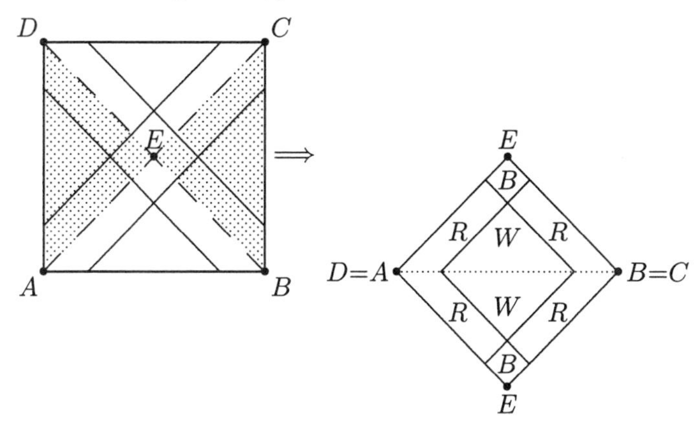

For simplicity assume that this "half-flag" is a 10×10 square. The interior white square consists of 64% of the area, so it must be 8×8. Thus the two blue squares measure 1×1, so they constitute 2% of the area.

22. **(A)** For a block to differ from the given block, there is only 1 choice for a different material, 2 choices for a different size, 3 choices for a different color, and 3 choices for a different shape. There are $\binom{4}{2} = 6$ ways a block can differ from the given block in exactly two ways:
    - **(1)** *Material and size:* $1 \cdot 2 = 2$ differing blocks.
    - **(2)** *Material and color:* $1 \cdot 3 = 3$ differing blocks.
    - **(3)** *Material and shape:* $1 \cdot 3 = 3$ differing blocks.
    - **(4)** *Size and color:* $2 \cdot 3 = 6$ differing blocks.
    - **(5)** *Size and shape:* $2 \cdot 3 = 6$ differing blocks.
    - **(6)** *Color and shape:* $3 \cdot 3 = 9$ differing blocks.

    Thus, $2 + 3 + 3 + 6 + 6 + 9 = 29$ blocks differ from the given block in exactly two ways.

    Note. The number of blocks that differ from the given block in exactly j ways is the coefficient of x^j in

$$(1 + x)(1 + 2x)(1 + 3x)(1 + 3x) = 1 + 9x + 29x^2 + 39x^3 + 18x^4.$$

This is an example of a *generating function*. [See almost any text on discrete mathematics or combinatorics.]

23. **(D)** Label the lattice points with the time t when the particle will achieve that position, and look for a pattern. Note that if t is even, after t^2 minutes all the lattice points of the square

$$0 \leq x \leq t-1, 0 \leq y \leq t-1$$

have been visited, and the particle is at $(0, t)$. Now,

$$1989 = 44^2+53, \text{ and } 53 = 44+9.$$

After $44^2 = 1936$ minutes the particle is at $(0, 44)$, 44 minutes later its location is $(44, 44)$, and 9 minutes after that it is at $(44, 35)$.

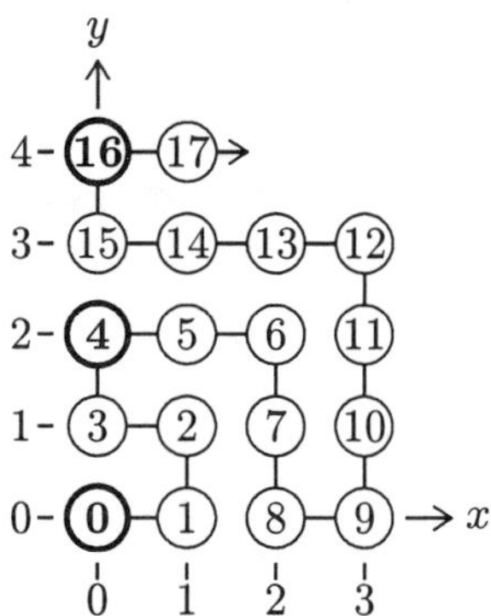

24. **(B)** If there are 0 females, then $(f, m) = (0, 5)$;
if there is 1 female, then $(f, m) = (2, 5)$;
if there are 2 females, then $(f, m) = (4, 5)$ or $(3, 4)$ depending on whether or not the females are sitting next to each other. This is easy to see using diagrams such as the following in which positions adjacent to an **F** are circled:

By symmetry, $(f, m) = (5, 0)$, $(f, m) = (5, 2)$, and $(f, m) = (5, 4)$ or $(4, 3)$ correspond to 5, 4 or 3 females, respectively. Thus there are $1 + 1 + 2 + 2 + 1 + 1 = 8$ possible (f, m) pairs.

25. **(B)** With exactly 10 runners contributing to their teams' scores, the sum of the scores of the two teams is

$$1 + 2 + 3 + \cdots + 10 = 55.$$

Consequently, every winning score must be less than half of 55. The lowest winning score is

$$1 + 2 + 3 + 4 + 5 = 15.$$

Thus, there could be 13 winning scores, $15, 16, 17, \ldots, 27$, provided that all these scores are possible.

In fact, all integers between 15 and 40 are possible scores. Note that if a certain finishing order results in score $x < 6 + 7 + 8 + 9 + 10$ for Team A, then there is a runner from Team A in that finishing order after whom the next finisher is from Team B. If the positions for these two runners were interchanged, then the resulting finishing order would give Team A a score of $x + 1$.

OR

Verify that every score from 15 through 27 is possible by listing one of the ways a team could obtain that score:

$$\begin{array}{ll}
1+2+3+4+5=15 & 1+2+3+6+10=22 \\
1+2+3+4+6=16 & 1+2+3+7+10=23 \\
1+2+3+4+7=17 & 1+2+3+8+10=24 \\
1+2+3+4+8=18 & 1+2+3+9+10=25 \\
1+2+3+4+9=19 & 1+2+4+9+10=26 \\
1+2+3+4+10=20 & 1+2+5+9+10=27 \\
1+2+3+5+10=21 &
\end{array}$$

26. **(C)** Without loss of generality, let the length of an edge of the cube be 2. Then the volume of the cube is 8 and each edge of the octahedron is $\sqrt{2}$. Bisect the octahedron into two pyramids with square bases. Each of the pyramids has altitude 1 and a base of area $(\sqrt{2})^2 = 2$, so the volume of each pyramid is $(1/3)(2)(1) = (2/3)$. The volume of the octahedron is thus $4/3$, so the required ratio is $(4/3)/8 = 1/6$.

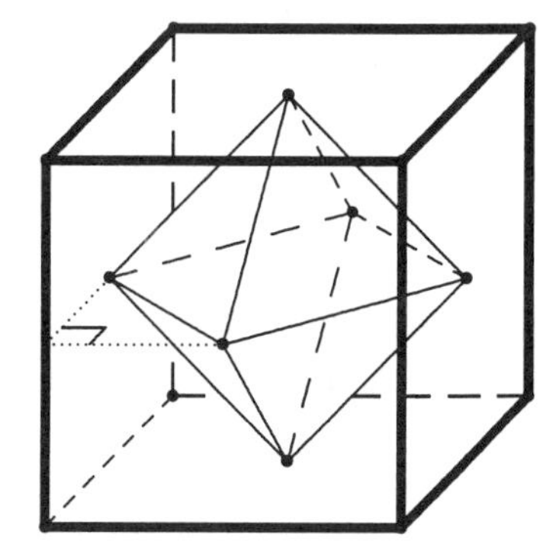

OR

The bases of the two pyramids that make up the octahedron each have area equal to half the area of a face of the cube. The height of each pyramid is half the height of the cube. Thus, where s is the length of an edge of the cube,

$$\frac{\text{volume of octahedron}}{\text{volume of cube}} = \frac{2\left(\frac{1}{3}\cdot\frac{s}{2}\cdot\frac{s^2}{2}\right)}{s^3} = \frac{1}{6}.$$

Note. More generally, if we form an octahedron by joining the centers of the six faces of <u>any</u> rectangular solid, then the volume of the octahedron is exactly one sixth of the volume of the rectangular solid.

Query. Is the hypothesis of perpendicularity in the preceding generalization a necessary one? Can you state and prove a further generalization?

27. **(D)** Because z is positive, solving $2(x+y) = n - z$ is equivalent to solving $x + y < \frac{n}{2}$ in positive integers x and y. The number of solutions to this inequality is the number of lattice points inside the triangle T in the first quadrant formed by the coordinate axes and the line $x + y = \frac{n}{2}$. [See figure which illustrates T for $n = 17$.] Since $1 + 2 + \cdots + 7 = 28$, n must be chosen so that there are exactly 7 lattice points on the line $y = 1$ in T. That is, $(1,7)$ must be inside triangle T, but $(1,8)$ must be on the boundary or outside of T. Hence $1 + 7 < \frac{n}{2} \le 1 + 8$, so that n is 17 or 18.

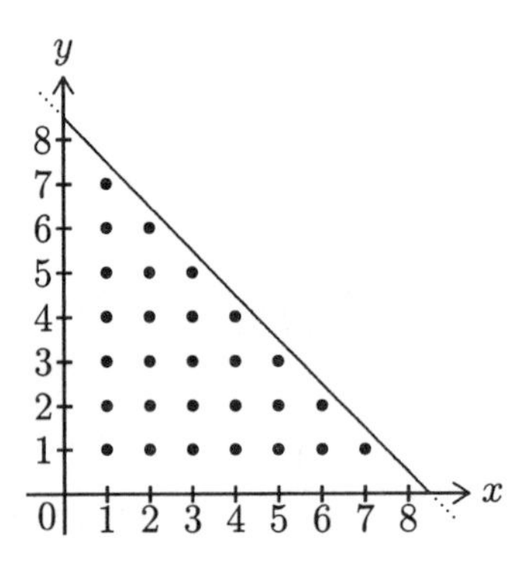

OR

Since $n - z = 2(x+y)$ must be even, we observe that n is even if and only if z is even.

Case 1: n and z both even. Then $z = 2j$ and $n = 2i$ for some positive integers i and j. Substitute in the original equation and divide by 2 throughout to obtain

$$x + y + j = i. \qquad (*)$$

Case 2: n and z both odd. Then $z = 2j - 1$ and $n = 2i - 1$ for some positive integers i and j. Substitute in the original equation and arrive at the same equation, $(*)$.

In each case $i = \lceil \frac{n}{2} \rceil$. To count the solutions to $(*)$, first note that since x and y are both positive integers,

$$j \in \{1, 2, 3, \ldots, i{-}2\}.$$

For each such j, there are $i - j - 1$ choices for (x, y):

$$(1, i{-}j{-}1),\ (2, i{-}j{-}2),\ (3, i{-}j{-}3),\ \ldots,\ (i{-}j{-}1, 1).$$

Summing over all possible j must yield 28:

$$\sum_{j=1}^{i-2}(i - j - 1) = \frac{(i-2)(i-1)}{2} = 28.$$

The positive solution is $i = 9$. Hence $\lceil \frac{n}{2} \rceil = 9$, so $n = 17$ or 18.

28. **(D)** Since

$$\begin{aligned}\tan^2 x - 9\tan x + 1 = \sec^2 x - 9\tan x &= \frac{1}{\cos^2 x} - 9\frac{\sin x}{\cos x}\\ &= \frac{1 - 9\sin x\cos x}{\cos^2 x} = \frac{1 - \frac{9}{2}\sin 2x}{\cos^2 x},\end{aligned}$$

we need to sum the roots of the equation $\sin 2x = \frac{2}{9}$ between $x = 0$ and $x = 2\pi$. These four roots are

$$x = \frac{\arcsin \frac{2}{9}}{2},\ \frac{\pi - \arcsin \frac{2}{9}}{2},\ \frac{2\pi + \arcsin \frac{2}{9}}{2},\ \text{and}\ \frac{3\pi - \arcsin \frac{2}{9}}{2}.$$

Since the $\arcsin \frac{2}{9}$ terms will cancel when we sum these four roots, their sum is

$$\frac{0}{2} + \frac{\pi}{2} + \frac{2\pi}{2} + \frac{3\pi}{2} = 3\pi.$$

OR

For any $b > 2$ the solutions of $y^2 - by + 1 = 0$ are

$$y_1, y_2 = \frac{b \pm \sqrt{b^2 - 4}}{2},$$

which are distinct and positive. Then either use the fact that the product of the roots of a quadratic is the constant term, or note that

$$y_1 \cdot y_2 = \frac{\left(b+\sqrt{b^2-4}\right)\left(b-\sqrt{b^2-4}\right)}{2^2} = \frac{b^2-(b^2-4)}{4} = 1$$

to see that y_1 and y_2 are reciprocals. Choose first-quadrant angles, x_1 and x_2, so $\tan x_1 = y_1$ and $\tan x_2 = y_2$. Then

$$\tan x_2 = y_2 = \frac{1}{y_1} = \frac{1}{\tan x_1} = \cot x_1 = \tan\left(\frac{\pi}{2} - x_1\right),$$

so x_1 and x_2 are complementary. Since $\tan(x+\pi) = \tan x$, there are four values of x between 0 and 2π, and all can be expressed interms of x_1:

$$x_1,\ \frac{\pi}{2} - x_1,\ \pi + x_1,\ \frac{3\pi}{2} - x_1.$$

Their sum is 3π.

29. **(B)** By the *Binomial Theorem*,

$$(1+i)^{99} = \binom{99}{0} + \binom{99}{1}i + \binom{99}{2}i^2 + \binom{99}{3}i^3 + \cdots + \binom{99}{99}i^{99}.$$

Note that the real part of this series is

$$\binom{99}{0} - \binom{99}{2} + \binom{99}{4} - \binom{99}{6} + \cdots - \binom{99}{98},$$

the very sum we want to find. Since

$$(1+i) = \sqrt{2}\left(\cos\frac{\pi}{4} + i\sin\frac{\pi}{4}\right),$$

by *DeMoivre's Theorem*

$$(1+i)^{99} = 2^{(99/2)}\left(\cos\frac{99\pi}{4} + i\sin\frac{99\pi}{4}\right).$$

Thus the real part of $(1+i)^{99}$ is

$$2^{99/2}\cos\frac{99\pi}{4} = 2^{99/2}\cos\frac{3\pi}{4} = 2^{99/2}\left(-\frac{\sqrt{2}}{2}\right) = -2^{49}.$$

30. **(A)** Suppose that John and Carol are two of the people. For $i = 1, 2, \ldots, 19$, let J_i and C_i be the numbers of orderings (out of all 20!)

in which the ith and $(i+1)$st persons are John and Carol, or Carol and John, respectively. Then $J_i = C_i = 18!$ is the number of orderings of the remaining persons.

For $i = 1, 2, \ldots, 19$, let N_i be the number of times a boy-girl or girl-boy pair occupies positions i and $i+1$. Since there are 7 boys and 13 girls, $N_i = 7 \cdot 13 \cdot (J_i + C_i)$. Thus the average value of S is

$$\frac{N_1 + N_2 + N_3 + \ldots + N_{19}}{20!} = \frac{19[7 \cdot 13 \cdot (18! + 18!)]}{20!} = \frac{91}{10}.$$

OR

In general, suppose there are k boys and $n-k$ girls. For $i = 1, 2, \ldots,$ $n-1$ let A_i be the probability that there is a boy-girl pair in positions $(i, i+1)$ in the line. Since there is either 0 or 1 pair in $(i, i+1)$, A_i is also the expected number of pairs in these positions. By symmetry, all A_i's are the same (or note that the argument below is independent of i). Thus the answer is $(n-1)A_i$.

We may consider the boys indistinguishable and likewise the girls. (*Why?*) Then an order is just a sequence of k **B**s and $n-k$ **G**s. To have a pair at $(i, i+1)$ we must have **BG** or **GB** in those positions, and the remaining $n-2$ positions must have $k-1$ boys and $n-k-1$ girls. Thus there are $2\binom{n-2}{k-1}$ sequences with a pair at $(i, i+1)$. Since there are $\binom{n}{k}$ sequences,

$$\textbf{answer} = (n-1)A_i = \frac{(n-1)2\binom{n-2}{k-1}}{\binom{n}{k}} = \frac{2k(n-k)}{n}.$$

Thus, when $n = 20$ and $k = 7$, the answer is $(2 \cdot 7 \cdot 13)/20 = 91/10$.

41 AHSME Solutions

1. **(E)** Multiply both sides of

$$\frac{x/4}{2} = \frac{4}{x/2} \quad \text{by} \quad 2 \cdot \frac{x}{2} \quad \text{to obtain} \quad \frac{x^2}{8} = 8.$$

Thus $x^2 = 8^2$, or $x = \pm 8$.

2. **(E)** $\left(\frac{1}{4}\right)^{-\frac{1}{4}} = \left(\frac{1}{2^2}\right)^{-\frac{1}{4}} = \left(2^{-2}\right)^{-\frac{1}{4}} = 2^{\frac{2}{4}} = 2^{\frac{1}{2}} = \sqrt{2}.$

OR

$$\left(\frac{1}{4}\right)^{-\frac{1}{4}} = 4^{1/4} = \sqrt[4]{4} = \sqrt{2}.$$

3. **(C)** Let d be the common difference of the arithmetic sequence. Then

$$75 + (75+d) + (75+2d) + (75+3d) = 360$$
$$300 + 6d = 360,$$

so $d = 10$ and the number of degrees in the largest angle is $75 + 3d = 105$.

OR

Since the opposite bases of a trapezoid are parallel, the smallest and largest angles must be supplementary. Consequently, the answer is $180° - 75° = 105°$.

Note. If four angles in an arithmetic sequence sum to $360°$, then the smallest and largest must be supplementary, as must the two intermediate sized angles. Therefore, any quadrilateral whose consecutive angles form an arithmetic sequence is a trapezoid. Any isosceles trapezoid shows that the converse is not true.

4. **(B)** Since $\overline{AD} \parallel \overline{BC}$, $\triangle FDE$ is similar to $\triangle BCE$. Hence

$$\frac{FD}{BC} = \frac{DE}{CE} \quad \text{or} \quad FD = \frac{DE}{CE} \cdot BC = \frac{4}{4+16} \cdot 10 = 2.$$

OR

Since $\angle BFA = \angle EFD$ and $\angle FAB = \angle FDE$, $\triangle FAB$ is similar to $\triangle FDE$. Hence

$$\frac{FA}{FD} = \frac{AB}{DE} \quad \text{or} \quad FA = \frac{AB \cdot FD}{DE} = \frac{16}{4} FD.$$

Since $FA + FD = AD = BC = 10$, we solve $4FD + FD = 10$ to find $FD = 2$.

Note. The answer is independent of $\angle ABC$.

5. **(B)** If a and b are positive, then $a < b$ if and only if $a^6 < b^6$. Since all the choices are positive, raise each to the sixth power to simplify the comparison:

(A) 5·6 **(B)** 6·6·6·5 **(C)** 5·5·5·6 **(D)** 5·5·6 **(E)** 6·6·5

and note that **(B)** is largest.

6. **(D)** The set of lines that are 2 units from the point A is the set of tangents to the circle with center A and radius 2. Similarly, the set of lines that are 3 units from point B is the set of tangents to the circle with center B and radius 3. Thus the desired set of lines is the set of common tangents to the two circles. Since $AB = 5 = 2+3$, these two circles are tangent externally, so they have three common tangents.

7. **(A)** Let $a \geq b \geq c$ be the lengths of the three sides of the triangle. The longest side of a triangle must be less than half the perimeter and at

least one third of the perimeter. Therefore, a is an integer that satisfies $8/2 > a \geq 8/3$, so $a = 3$. Since

$$3 \geq b \geq c > 0 \text{ and } b + c = 8 - a = 5,$$

the only triangle with integral sides and perimeter 8 has sides of lengths 3, 3 and 2. The altitude to the base of length 2 of this isosceles triangle is $\sqrt{3^2 - 1^2}$, so its area is

$$\frac{1}{2} \cdot 2 \cdot \sqrt{8} = 2\sqrt{2}.$$

Query. If any integral perimeter larger than 8 were used, the area would not be unique. Can you prove it?

8. **(E)** For any number x in the interval $[2, 3]$, we have

$$|x - 2| + |x - 3| = (x - 2) + (3 - x) = 1.$$

OR

We can interpret $|a-b|$ as the distance between a and b. Then the given equation is the condition that the distance between x and 2 plus the distance between x and 3 equals 1. This is the case whenever $2 \leq x \leq 3$.

OR

Plot $y = |x - 2|$ and $y = |x - 3|$ on the same axes. Then for each x, add the y-coordinates on these graphs to find the y coordinate on the graph of

$$y = |x-2| + |x-3|.$$

Note that there is an interval of points on this graph at height $y = 1$.

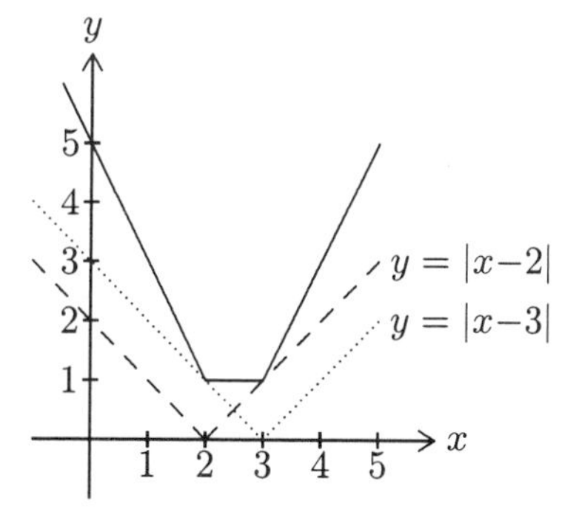

9. **(B)** Since there are six faces and each edge is shared by only two faces, there must be at least $6/2 = 3$ black edges. The diagram shows that 3 black edges will suffice.

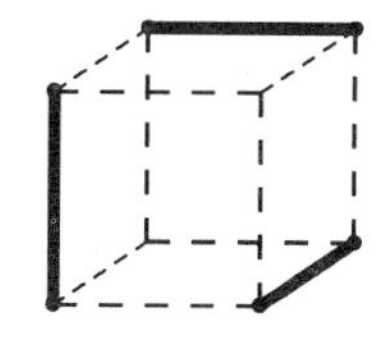

10. **(D)** At most three of the large cube's six faces can be seen at once. Excluding the unit cubes of the three closest edges, the three visible faces contain $3\cdot 10^2$ unit cubes. The three edges contain $3\cdot 10$ unit cubes plus the single, shared corner cube. Therefore the desired number is $3\cdot 10^2+3\cdot 10+1=331$.

OR

The unseen unit cubes form a $10\times 10\times 10$ cube. Thus the number of unit cubes that can be seen is $11^3-10^3=331$.

OR

In one of the three visible faces of the large cube we can see 11^2 unit cubes, on another we can see 11^2-11 which were not seen on the first side, and on the remaining face we can see $(11-1)^2$ which were not seen on either of the first two faces. Thus the desired number is $11^2+(11^2-11)+(11-1)^2=331$.

11. **(C)** Let N be a positive integer and d a divisor of N. Then N/d is also a divisor of N. Thus the divisors of N occur in pairs $d, N/d$ and these two divisors will be distinct unless N is a perfect square and $d=\sqrt{N}$. It follows that N has an odd number of divisors if and only if N is a perfect square. There are 7 perfect squares, $1^2, 2^2, 3^2, 4^2, 5^2, 6^2, 7^2$, among the numbers $1, 2, 3, \ldots, 50$.

Note. If $N>1$ is an integer then

$$N=p_1^{r_1}\cdot p_2^{r_2}\cdot\ldots\cdot p_k^{r_k}$$

where p_i is the ith prime. The divisors of N are those

$$d=p_1^{s_1}\cdot p_2^{s_2}\cdot\ldots\cdot p_k^{s_k}$$

with $0\le s_i\le r_i$ for all i. Thus, N has

$$(r_1+1)\cdot(r_2+1)\cdot\ldots\cdot(r_k+1)$$

divisors, a product which will be an odd number only when each r_i is even. Each r_i is even if and only if N is a perfect square.

12. **(D)** Since $a\neq 0$, the only x for which $f(x)=-\sqrt{2}$ is $x=0$. Since $f(f(\sqrt{2}))=-\sqrt{2}$, $f(\sqrt{2})$ must be 0. Finally,

$$f\left(\sqrt{2}\right)=0 \iff a\left(\sqrt{2}\right)^2-\sqrt{2}=0 \iff a=\frac{\sqrt{2}}{2}.$$

OR

Since $f(\sqrt{2}) = 2a - \sqrt{2}$,

$$f\left(f(\sqrt{2})\right) = a(2a - \sqrt{2})^2 - \sqrt{2}$$

which we set equal to $-\sqrt{2}$. Therefore, $a(2a-\sqrt{2})^2 = 0$. Since $a > 0$, we have $2a = \sqrt{2}$ and thus $a = \sqrt{2}/2$.

13. **(E)** Since $S = 5 + 7 + 9 + \cdots + X$ is the sum of an arithmetic series containing

$$1 + \frac{X-5}{2}$$

terms, we have

$$S = \frac{1}{2}\left(1 + \frac{X-5}{2}\right)(5 + X) = \frac{1}{2}\left(\frac{X-3}{2}\right)(X+5)$$
$$= \frac{X^2 + 2X - 15}{4} = \left(\frac{X+1}{2}\right)^2 - 4 \geq 10000.$$

Thus $(X+1)/2 \geq \sqrt{10004}$. Since X is an odd integer, $X = 201$ is the printed value.

OR

Since X is large, approximate

$$\frac{X^2 + 2X - 15}{4} \geq 10000 \text{ with } \frac{X^2}{4} > 100^2$$

and test odd integers near 200 to find $X = 201$.

14. **(A)** Angles BAC, BCD and CBD all intercept the same circular arc, minor arc BC of measure $2x$ since $\angle A = x$. Therefore, $\angle BCD = \angle CBD = x$ and, considering the sum of the angles in $\triangle BCD$, $\angle D = \pi - 2x$. Since $\angle ABC = \angle ACB$, considering the sum of the angles in $\triangle ABC$ we have $\angle ABC = (\pi - x)/2$. The given condition, $\angle ABC = 2\angle D$, now becomes

$$\frac{\pi - x}{2} = 2(\pi - 2x), \quad \text{so} \quad x = \frac{3\pi}{7}.$$

OR

Let O be the center of the circle. Since $\triangle OAB$ is isosceles, $\angle ABO = \angle BAO = x/2$. Since $\overline{BD}$ is tangent to the circle, $\angle OBD = \pi/2$. Thus

$$\angle ABD = \angle ABO + \angle OBD = \frac{x}{2} + \frac{\pi}{2}.$$

Since $\triangle ABC$ is isosceles,

$$\angle ABC = \frac{\pi - x}{2}, \text{ so}$$

$$\angle D = \frac{1}{2}\angle ABC = \frac{\pi - x}{4}.$$

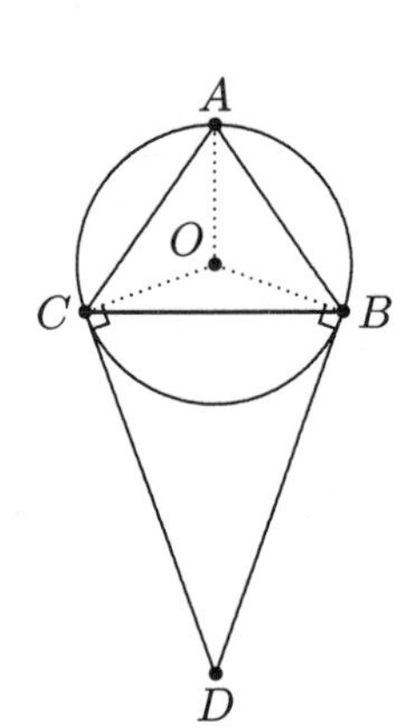

Sum the angles of quadrilateral $ABDC$ to obtain

$$\begin{aligned} 2\pi &= \angle CAB + \angle ABD + \angle D + \angle DCA \\ &= x + \left(\frac{x}{2} + \frac{\pi}{2}\right) + \frac{\pi - x}{4} + \left(\frac{x}{2} + \frac{\pi}{2}\right) \end{aligned}$$

which we solve to obtain $x = 3\pi/7$.

OR

Let O be the center of the circle. Then $\angle COB = 2x$ and, from the sum of the angles of the quadrilateral $COBD$, we obtain $2x + \angle D = \pi$. The conditions of the problem yield $x + 4\angle D = \pi$ to be the sum of the angles of $\triangle ABC$. Solve these two equations in x and $\angle D$ simultaneously to find $x = 3\pi/7$.

Query. What is x if $\triangle ABC$ is an obtuse isosceles triangle?

15. **(C)** Let the four numbers be w, x, y and z with $w \le x \le y \le z$. Then

$$\begin{array}{rrcl} & w + x + y \phantom{{}+z} & = & 180 \\ & w + x \phantom{{}+y} + z & = & 197 \\ & w \phantom{{}+x} + y + z & = & 208 \\ & x + y + z & = & +222 \\ \hline \text{so} & 3(w + x + y + z) & = & 807 \\ \text{and hence} & w + x + y + z & = & 269. \end{array}$$

Since $w + x + y + z = 269$ and $w + x + y = 180$, we have $z = 269 - 180 = 89$.

OR

Simply note that since each number appears three times in the four sums,

$$3(w + x + y + z) = 180 + 197 + 208 + 222 = 807.$$

and conclude as above.

16. **(C)** If all 26 people shook hands there would be $\binom{26}{2}$ handshakes. Of these, $\binom{13}{2}$ would take place between women and 13 between spouses. Therefore there were

$$\binom{26}{2} - \binom{13}{2} - 13 = 13 \cdot 25 - 13 \cdot 6 - 13 = 234$$

handshakes.

OR

For a party attended by n couples, the number of handshakes between men is $\frac{1}{2}n(n-1)$ and the number of handshakes between unmarried persons of different genders is $n(n-1)$. Therefore there is a total of

$$\frac{1}{2}n(n-1) + n(n-1) = \frac{3}{2}n(n-1)$$

handshakes. When $n = 13$ there are $(3 \cdot 13 \cdot 12)/2 = 234$ handshakes.

17. **(C)** For every 3 distinct digits selected from $\{1, 2, \ldots, 9\}$ there is exactly one way to arrange them into a number with increasing digits, and every number with increasing digits corresponds to one of these selections. Similarly, the numbers with decreasing digits correspond to the subsets with 3 elements of the set of all 10 digits. Hence our answer is

$$\begin{aligned}\binom{9}{3} + \binom{10}{3} &= \frac{9 \cdot 8 \cdot 7}{1 \cdot 2 \cdot 3} + \frac{10 \cdot 9 \cdot 8}{1 \cdot 2 \cdot 3} \\ &= \frac{9 \cdot 8}{2 \cdot 3}(7 + 10) = 12 \cdot 17 = 204.\end{aligned}$$

OR

Make a list in decreasing order of the three-digit numbers with increasing digits, grouping them according to their first and second digits:

$$789,$$

$$689, \underbrace{679, 678,}_{2}$$

$$589, \underbrace{579, 578,}_{2} \underbrace{569, 568, 567,}_{3}$$

$$\vdots$$

$$189, \underbrace{179, 178,}_{2} \ldots, \underbrace{129, 128, \ldots, 123}_{7}$$

There are

$$1 + (1+2) + (1+2+3) + \cdots + (1+2+\cdots+7) = 84$$

such numbers.

Now make a list in increasing order of the three-digit numbers with decreasing digits, grouping them by first and second digits:

$$210,$$

$$310, \underbrace{320, 321,}_{2}$$

$$410, \underbrace{420, 421,}_{2} \underbrace{430, 431, 432,}_{3}$$

$$\vdots$$

$$910, \underbrace{920, 921,}_{2} \ldots, \underbrace{980, 981, \ldots, 987}_{8}$$

In this case there are

$$1 + (1{+}2) + (1{+}2{+}3) + \cdots + (1{+}2{+}\cdots{+}8) = 120$$

such numbers, an additional $(1{+}2{+}\cdots{+}8) = 36$ numbers since a number with decreasing digits can contain 0 while one with increasing digits cannot. Thus, the answer is $84 + 120 = 204$.

18. **(C)** Observe the repeating pattern of the units digits of consecutive integral powers of 3 and 7:

$$\begin{array}{ll} 3^1 = \mathbf{3} & 7^1 = \mathbf{7} \\ 3^2 = \mathbf{9} & 7^2 = 4\mathbf{9} \\ 3^3 = 2\mathbf{7} & 7^3 = 34\mathbf{3} \\ 3^4 = 8\mathbf{1} & 7^4 = 240\mathbf{1} \\ 3^5 = 24\mathbf{3} & 7^5 = 1680\mathbf{7} \end{array}$$

Note that 25 of the given values for a yield a units digit in 3^a of 3, 25 yield 9, 25 yield 7 and 25 yield 1. Similarly, the given values for b yield 25 of each of these units digits in 7^b: 7, 9, 3, 1. Thus there are 16 possible pairs of units digits, and each pair is equally likely:

$$\begin{array}{cccc} (3,7) & (3,9) & (3,3) & (3,1) \\ (9,7) & \boxed{(\mathbf{9},\mathbf{9})} & (9,3) & (9,1) \\ (7,7) & (7,9) & (7,3) & \boxed{(\mathbf{7},\mathbf{1})} \\ \boxed{(\mathbf{1},\mathbf{7})} & (1,9) & (1,3) & (1,1) \end{array}$$

Of these, three pairs, $(1,7)$, $(7,1)$ and $(9,9)$, yield a sum with units digit 8. Thus, the desired probability is $3/16$.

19. **(B)** Since

$$\frac{N^2+7}{N+4} = \frac{(N-4)(N+4)+23}{N+4} = N-4+\frac{23}{N+4},$$

the numerator and denominator will have a nontrivial common factor exactly when $N+4$ and 23 have a factor in common. Because 23 is a prime, $N+4$ is a multiple of 23 when $N = -4+23k$ for some integer k. Solving

$$1 < -4+23k < 1990 \quad \text{yields} \quad \frac{5}{23} < k < 86\frac{16}{23},$$

or $k = 1, 2, \ldots, 86$.

Note. If $b = aq + r$, or, equivalently,

$$\frac{b}{a} = q + \frac{r}{a},$$

then the greatest common factor of a and b will be the same as the greatest common divisor of r and a. This is basis of the Euclidean Algorithm. Ordinarily we might think of this fact when a, b, q and r are integers, but it also holds when they are polynomial expressions.

20. **(C)** Since $\angle BAF$ and $\angle ADE$ are both complementary to $\angle CAD$, they must be equal. Thus, $\triangle BAF \sim \triangle ADE$ so

$$\frac{BF}{AE} = \frac{AF}{DE}, \quad \text{or} \quad \frac{BF}{3} = \frac{3+EF}{5}.$$

By an analogous argument, $\triangle BCF \sim \triangle CDE$,

$$\frac{BF}{CE} = \frac{CF}{DE}, \quad \text{and} \quad \frac{BF}{7} = \frac{7-EF}{5}.$$

Solve

$$5BF - 3EF = 9$$

$$5BF + 7EF = 49$$

simultaneously to obtain $BF = 4.2$.

OR

Note that $ABCD$ is a cyclic quadrilateral since opposite angles are supplementary. Extend $\overline{DE}$ to X on the circumcircle. Since $\angle DAB$ subtends the same arc as $\angle DXB$, $BFEX$ is a rectangle and $BF = EX$. We can consider $\overline{AC}$ and $\overline{DX}$ as intersecting chords in a circle and use $DE \cdot EX = AE \cdot EC$ to find that

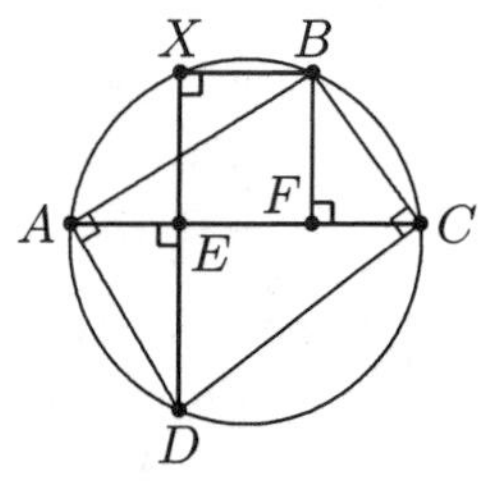

$$BF = EX = \frac{AE \cdot EC}{DE} = \frac{3 \cdot 7}{5} = \frac{21}{5}.$$

OR

Angles ABF and EAD are equal since they are complementary to $\angle BAC$. Similarly, $\angle BCF = \angle CDE$. Thus

$$\begin{aligned} 10 = AC = AF + FC &= BF \tan \angle ABF + BF \cot \angle BCF \\ &= BF \tan \angle EAD + BF \cot \angle CDE \\ &= BF \cdot \frac{5}{3} + BF \cdot \frac{5}{7} = BF \cdot \frac{50}{21}. \end{aligned}$$

Hence $BF = 10 \cdot (21/50) = 21/5 = 4.2$.

OR

By the Pythagorean Theorem,

$$\begin{aligned}(BF^2+FA^2)+(AE^2+ED^2) &= BA^2+AD^2\\ &= BD^2\\ &= BC^2+CD^2\\ &= (BF^2+FC^2)+(DE^2+EC^2).\end{aligned}$$

Subtract like terms to obtain $AF^2 - FC^2 = CE^2 - EA^2$, which we can rewrite as

$$(AF+FC)(AF-FC) = (CE+EA)(CE-EA)$$

$$(AC)\big((AC-CF)-CF\big) = (AC)\big((AC-AE)-AE\big)$$

$$(AC)(AC-2CF) = (AC)(AC-2AE).$$

Since $AC \neq 0$, it follows that $CF = AE = 3$, and therefore $EF = 4$ and $CE = 7$.

Argue as above that $\triangle BCF \sim \triangle CDE$. Therefore

$$\frac{BF}{CE} = \frac{CF}{DE} \quad \text{or} \quad BF = \frac{CE\cdot CF}{DE} = \frac{7\cdot 3}{5} = 4.2.$$

21. **(E)** Let M be the midpoint of $\overline{AB}$ and O be the center of the square. Thus $AM = OM = \frac{1}{2}$ and slant height $PM = \frac{1}{2}\cot\theta$. Hence

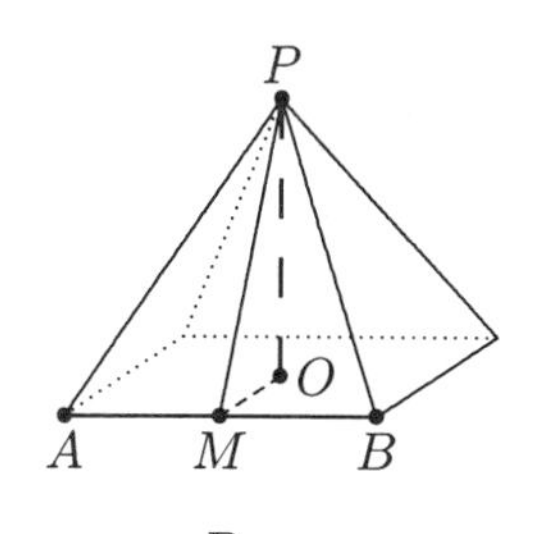

$$\begin{aligned}PO^2 = PM^2 - OM^2 &= \frac{1}{4}\cot^2\theta - \frac{1}{4}\\ &= \frac{\cos^2\theta - \sin^2\theta}{4\sin^2\theta} = \frac{\cos 2\theta}{4\sin^2\theta}.\end{aligned}$$

Since $0 < \theta < 45°$, the volume is

$$\frac{1}{3}\cdot 1^2\cdot PO = \frac{\sqrt{\cos 2\theta}}{6\sin\theta}.$$

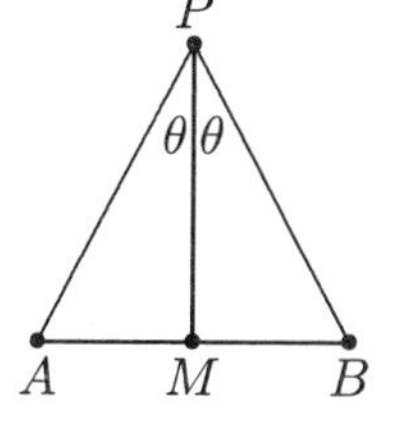

22. **(D)** Use the polar form, $x = r(\cos\theta + i\sin\theta)$. By DeMoivre's Theorem,

$$r^6(\cos 6\theta + i\sin 6\theta) = x^6 = -64 =$$

$$2^6\big(\cos(180° + 360°k) + i\sin(180° + 360°k)\big).$$

Thus $r = 2$ and, using $k = -3, -2, -1, 0, 1, 2$, we have

$$\theta = (180^\circ + 360^\circ k)/6 = \pm 30^\circ, \pm 90^\circ, \pm 150^\circ.$$

Since $x = r(\cos\theta + i\sin\theta) = a + bi$ and $a > 0$, it follows that $\theta = \pm 30^\circ$ are the only values to consider. Thus

$$x = 2\big(\cos(\pm 30^\circ) + i\sin(\pm 30^\circ)\big) = \sqrt{3} \pm i.$$

The product of these two roots is

$$\left(\sqrt{3} + i\right)\left(\sqrt{3} - i\right) = 4.$$

OR

Recall, from DeMoivre's Theorem, that the six sixth roots of -64 are equispaced around the circle of radius $\sqrt[6]{64}$. Since $\pm 2i$ are roots, exactly two of the roots are in the right half-plane and they must be conjugates. The product of any pair of conjugates is the square of their distance from the origin, so the product of these two roots is $\left(\sqrt[6]{64}\right)^2 = 4$.

OR

Since the equation has real coefficients, complex roots occur in conjugate pairs. Furthermore, $a \pm bi$, $a > 0$ are roots of $x^6 + 64 = 0$ if and only if $x^2 - 2ax + a^2 + b^2$ is a factor of $x^6 + 64$. Now

$$\begin{aligned} x^6 + 64 &= (x^2 + 4)(x^4 - 4x^2 + 16) \\ &= (x^2 + 4)\big((x^2 + 4)^2 - 12x^2\big) \\ &= (x^2 + 4)(x^2 - 2\sqrt{3}x + 4)(x^2 + 2\sqrt{3}x + 4). \end{aligned}$$

The desired roots occur in the middle factor since that is the only one for which the coefficient of x is negative. The product of the roots of that factor is its constant term, 4.

OR

The equation $x^6 + 64 = 0$ has real coefficients, so the complex roots occur in conjugate pairs. The coefficient of x^5 is 0, so the sum of the roots is 0. Since two of the roots are $\pm 2i$ and the roots must occur in conjugate pairs, the other four roots must be $a+bi$, $a-bi$, $-a+bi$ and $-a-bi$ for some $a>0$ and $b>0$ and the answer will be

$$(a+bi)(a-bi)=a^2+b^2.$$

The product of all six roots must the the constant term in the equation, 64, so

$$\begin{aligned}(2i)(-2i)(a+bi)(a-bi)(-a+bi)(-a-bi)&=64\\ 4(a^2+b^2)^2&=64,\end{aligned}$$

so $a^2+b^2=\sqrt{64/4}=4$.

23. **(B)** By symmetry in x and y, we can assume that $x\geq y$, and therefore that $\log_y x\geq 1$. Let $v=\log_y x$. Then, since $\log_x y=1/v$ solve

$$v+\frac{1}{v}=\frac{10}{3}$$

to find $v=3$ or $v=1/3$. Since $v=\log_y x\geq 1$, $v=3$, so

$$\begin{aligned}\log_y x&=3\\ x&=y^3\\ y^4=(y^3)y=xy=144&=\left(\sqrt{12}\right)^4\\ y=\sqrt{12}&=2\sqrt{3}\quad\text{and}\quad x=y^3=24\sqrt{3}.\end{aligned}$$

Thus

$$\frac{x+y}{2}=\frac{24\sqrt{3}+2\sqrt{3}}{2}=13\sqrt{3}.$$

24. **(D)** We want a weighted average, X, of 76 and 90, with weights proportional to the number of girls at Adams HS and Baker HS, respectively. We obtain these weights as follows: Let

b=number of boys at Adams, B=number of boys at Baker,
g=number of girls at Adams, G=number of girls at Baker.

From the first column of the table obtain

$$\frac{71b+76g}{b+g}=74,$$

so we solve $71b + 76g = 74b + 74g$ to find that $g = 1.5b$. Similarly, the second column shows that $G = .5B$ and the first row shows that $B = 4b$. Thus

$$X = \frac{76g + 90G}{g + G} = \frac{76(1.5b) + 90[.5(4b)]}{1.5b + [.5(4b)]} = \frac{114 + 180}{3.5} = 84.$$

25. **(B)** Let r be the radius of each sphere, and let O be the center of the cube. Consider the rectangular cross section $ABCD$ of the cube through O where $\overline{AD}$ and $\overline{BC}$ are parallel edges of the cube that join the parallel face diagonals $\overline{AB}$ and $\overline{CD}$. Since $\overline{AC}$ is one of the interior diagonals of the cube, the centers of the spheres E, O and F lie on $\overline{AC}$ and $AC = \sqrt{3}$. Since $\triangle FGC \sim \triangle ADC$,

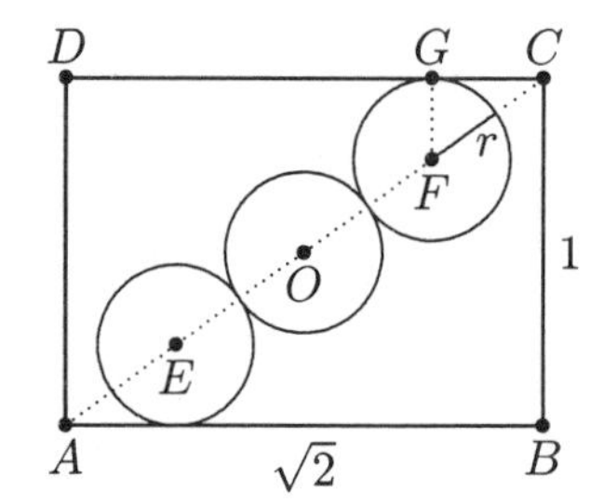

$$\frac{r}{1} = \frac{FG}{AD} = \frac{FC}{AC} = \frac{FC}{\sqrt{3}},$$

from which it follows that $FC = r\sqrt{3}$. Since $AE = FC$ and $EF = 4r$, we have

$$r\sqrt{3} + 4r + r\sqrt{3} = AE + EF + FC = AC = \sqrt{3}.$$

Solve this equation to find

$$r = \frac{\sqrt{3}}{4 + 2\sqrt{3}}\left(\frac{4 - 2\sqrt{3}}{4 - 2\sqrt{3}}\right) = \frac{4\sqrt{3} - 6}{4} = \sqrt{3} - \frac{3}{2}.$$

OR

If the radius of each sphere is r, the center of a corner sphere is $\sqrt{r^2 + r^2 + r^2}$ units from the closest vertex. From this computation of $FC = r\sqrt{3}$, continue the computations as above.

OR

Let r be the radius of each sphere. Note that the centers of the eight outer spheres form a cube of side $(1 - 2r)$ whose main diagonal is $4r$ units. Since the length of the diagonal of a cube is $\sqrt{3}$ times its side, $\sqrt{3}(1 - 2r) = 4r$. Solve this equation to find $r = \dfrac{2\sqrt{3} - 3}{2}$.

Note. To visualize the arrangement of the spheres in the cube, begin with nine small congruent spheres with one at the center and one tangent to the three faces at each of the eight vertices of the cube. Keeping the center sphere in the center of the cube and the other eight tangent to their three faces, expand the radii of all nine spheres until the spheres are tangent.

26. **(A)** Let p_i be the person who announced "i" and let x be the number picked by p_6. Since the average of the numbers picked by p_4 and p_6 is 5, p_4 picked $10-x$. Continuing counterclockwise around the table, we find that p_2 picked $x-4$, p_{10} picked $6-x$, p_8 picked $12+x$, and p_6 picked $2-x$. Since $2-x=x$, $x=1$.

OR

Let x_i be the original number picked by the person who announced "i". We have a system of ten equations in ten unknowns which has a unique solution:

$$\frac{1}{2}(x_{10}+x_2)=1, \qquad \frac{1}{2}(x_1+x_3)=2,$$
$$\frac{1}{2}(x_2+x_4)=3, \qquad \frac{1}{2}(x_3+x_5)=4,$$
$$\vdots \qquad\qquad \vdots$$
$$\frac{1}{2}(x_8+x_{10})=9, \qquad \frac{1}{2}(x_9+x_1)=10.$$

The sum of the five equations involving the variables with even subscripts yields $x_2+x_4+x_6+x_8+x_{10}=25$. Substitute $x_2+x_4=6$ and $x_8+x_{10}=18$ to obtain $x_6=1$. In the figure we show "i":$\mathbf{x}_i$ where the $\mathbf{x}_i$ yield the desired averages, "i".

"1":**6**

"10":**5** "2":**− 3**

"9":**14** "3":**− 2**

"8":**13** "4":**9**

"7":**2** "5":**10**

"6":**1**

Query. Suppose there had been n people instead of 10. For which n is there a unique answer?

Note. This problem is an example of inverting averages. Such problems arise in many applications of mathematics, for instance, the operation of CAT scanners in medicine. To obtain information from a CAT scan, one must invert averages along continuous rays in a disk, rather than averages of discrete points on the perimeter of the disk.

27. **(C)** Let x, y and z denote the sides of a triangle, h_x, h_y and h_z the corresponding altitudes, and A the area. Since $xh_x = yh_y = zh_z = 2A$, the sides are inversely proportional to the altitudes. If x, y and z form a triangle with largest side x, then $x < y + z$. Thus

$$\frac{2A}{h_x} < \frac{2A}{h_y} + \frac{2A}{h_z} \qquad \text{or} \qquad \frac{1}{h_x} < \frac{1}{h_y} + \frac{1}{h_z}. \tag{$*$}$$

Only triple **(C)** fails to satisfy $(*)$:

(A) $\frac{1}{2} < \frac{1}{\sqrt{3}} + \frac{1}{1}$ **(B)** $\frac{1}{3} < \frac{1}{4} + \frac{1}{5}$ **(C)** $\frac{1}{5} \not< \frac{1}{12} + \frac{1}{13}$
(D) $\frac{1}{7} < \frac{1}{8} + \frac{1}{\sqrt{113}}$ **(E)** $\frac{1}{8} < \frac{1}{15} + \frac{1}{17}$

To show that the other four choices (h_x, h_y, h_z) *do* correspond to possible triangles, just build a triangle T with sides $1/h_x$, $1/h_y$ and $1/h_z$. The altitudes of T are in the ratio $h_x{:}h_y{:}h_z$, so some triangle similar to T has altitudes h_x, h_y and h_z.

Note. Of the five inequalities above, **(A)**, **(B)**, **(D)** are immediate since $h_y, h_z < 2h_x$. In **(C)**, $h_y, h_z > 2h_x$. For **(E)** we have

$$\frac{1}{15} + \frac{1}{17} = \frac{15+17}{15 \cdot 17} = \frac{32}{(16-1)(16+1)} > \frac{32}{16^2} = \frac{1}{8}.$$

28. **(B)** Let the inscribed circle have center O and radius r. Label the quadrilateral $ABCD$ where $DA = 90$, $AB = 130$ and $BC = 110$. Label the points of tangency with the inscribed circle E, F, G and H, and let w, x, y and z be the distances from these points of tangency to the vertices of the quadrilateral as indicated in the figure. Since the quadrilateral is inscribed in a circle, $\angle DAB$ is supplementary to $\angle DCB$. Since $\overline{OA}$ bisects $\angle DAB$ and $\overline{OC}$ bisects $\angle DCB$, $\angle OAE$ and $\angle OCG$ are complementary. Hence $\triangle OEA \sim \triangle CGO$. Thus $x/r = r/z$. Similarly, $y/r = r/w$. Hence $wy = r^2 = zx$, which leads to

$$90y = (w + x)y = wy + xy = zx + xy = (z + y)x = 110x.$$

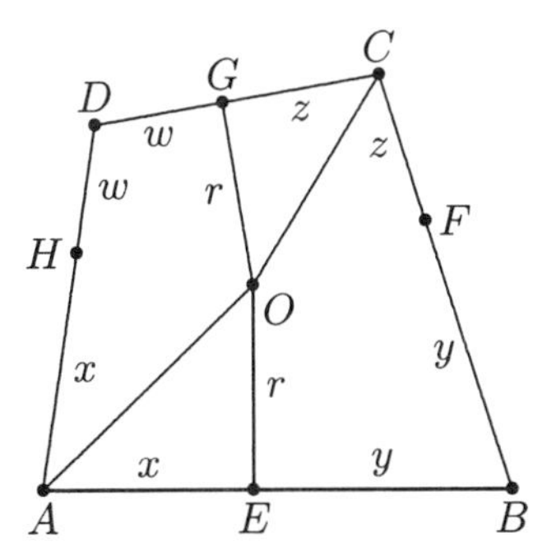

Solve $90y = 110x$ and $x + y = 130$ simultaneously to find that $x = (13 \cdot 9)/2$ and $y = (13 \cdot 11)/2$. Hence, $|x - y| = 13$.

Note. This quadrilateral has an inscribed circle because $70 + 130 = 90 + 110$. The shape of this quadrilateral is unique since it is inscribed in a circle.

29. **(D)** For each positive integer b that is not divisible by 3, we must decide which of the numbers in the list $b, 3b, 9b, 27b, \cdots \leq 100$ to place in the subset. Clearly, a maximal subset can be obtained by using alternate numbers from this list starting with b. Thus, it will contain $67 = 100 - 33$ members b that are not divisible by 3, $8 = 11-3$ members of the form $9b$ that are divisible by 9 but not by 27, and the number 81, for a total of $67 + 8 + 1 = 76$ elements.

OR

Let S be the set of all integers 1 through 100 of the form $a3^i$, i even and a relatively prime to 3. No element of S is 3 times another. Since the number of integers, $1, 2, \ldots, 100$, of the form $a3^i$ is $\left\lfloor \frac{100}{3^i} \right\rfloor$, S contains

$$100 - \left\lfloor \frac{100}{3} \right\rfloor + \left\lfloor \frac{100}{9} \right\rfloor - \left\lfloor \frac{100}{27} \right\rfloor + \left\lfloor \frac{100}{81} \right\rfloor$$

$= 100 - 33 + 11 - 3 + 1 = 76$ elements. Hence the answer is at least 76.

There are 24 pairs $(n, 3n)$ with $3n \leq 100$, $n = a3^i$, i=0 or 2, and a relatively prime to 3. Both n and $3n$ cannot be in a subset required by our problem. Thus the answer is at most $100 - 24 = 76$.

Query. If $S \subset \{1, 2, \ldots, n\}$ and no element of S is k times another, it appears that the largest possible number of elements in S is $n - \left\lfloor \frac{n}{k} \right\rfloor + \left\lfloor \frac{n}{k^2} \right\rfloor - \left\lfloor \frac{n}{k^3} \right\rfloor \cdots ? \sum_{i=0}^{+\infty} (-1)^i \left\lfloor \frac{n}{k^i} \right\rfloor$. Is this in fact the case for all integers $k > 1$?

Note. The maximal subset is not unique. For example, for each b between 13 and 32 that is not divisible by 3, either b or $3b$ could be used.

30. **(E)** Multiply both sides of $R_n = \dfrac{a^n + b^n}{2}$ by $a+b$ to obtain

$$\begin{aligned}(a+b)R_n &= (a+b)\left(\frac{a^n + b^n}{2}\right) \\ &= \frac{a^{n+1} + b^{n+1}}{2} + ab\left(\frac{a^{n-1} + b^{n-1}}{2}\right) \\ &= R_{n+1} + abR_{n-1}.\end{aligned}$$

Since $a+b = 6$ and $ab = 1$, the recursion

$$R_{n+1} = 6R_n - R_{n-1}$$

follows. Use this, together with $R_0 = 1$ and $R_1 = 3$, to calculate the units digits of

$$R_2, R_3, R_4, R_5, R_6, R_7, \ldots$$

which are $7, 9, 7, 3, 1, 3, 7, 9, \ldots$, respectively. An induction argument shows that R_n and R_{n+6} have the same units digit for all nonnegative n. In particular,

$$R_3, R_9, \cdots, R_{12345}$$

all have the same units digit, 9, since $12345 = 3 + 6 \cdot 2057$.

OR

We may compute a few powers of a and b to try to detect a pattern:

$$\begin{array}{ll} a^0 = 1 + 0\sqrt{2} & b^0 = 1 - 0\sqrt{2} \\ a^1 = 3 + 2\sqrt{2} & b^1 = 3 - 2\sqrt{2} \\ a^2 = 17 + 12\sqrt{2} & b^2 = 17 - 12\sqrt{2} \\ a^3 = 99 + 70\sqrt{2} & b^3 = 99 - 70\sqrt{2} \\ \vdots & \vdots \\ a^n = c_n + d_n\sqrt{2} & b^n = c_n - d_n\sqrt{2} \end{array}$$

Eventually one may notice that

$$c_n = 3c_{n-1} + 4d_{n-1}$$
$$d_n = 2c_{n-1} + 3d_{n-1}$$

These relations can be verified by mathematical induction. Since the units digit of c_n and d_n depend only on the units digits of c_{n-1} and d_{n-1}, we may use these formulas to list the units digits of c_n and d_n:

n	units digit of c_n	units digit of d_n
0	1	0
1	3	2
2	7	2
3	9	0
4	7	8
5	3	8
6	1	0

The answer is the units of digit of c_{12345}, which we can see is the same as the units digit of c_3 by the cyclic nature of the table.

Note. Since $(x-a)(x-b) = (x-3-2\sqrt{2})(x-3+2\sqrt{2}) = x^2-6x+1$ it follows that a and b satisfy $x^2 = 6x - 1$, so $a^{n+1} = a^{n-1}a^2 = a^{n-1}(6a-1) = 6a^n - a^{n-1}$ and similarly, $b^{n+1} = 6b^n - b^{n-1}$. This yields an alternate derivation of the recursion:

$$\begin{aligned} R_{n+1} &= \frac{a^{n+1} + b^{n+1}}{2} \\ &= 6\frac{a^n + b^n}{2} - \frac{a^{n-1} + b^{n-1}}{2} = 6R_n - R_{n-1}. \end{aligned}$$

Query. Here $R_{n+6} - R_n$ is divisible by 10. Further investigation for small i suggests that $R_{5^{i-1}(n+6)} - R_n$ is divisible by 10^i for $i \geq 1$. Can you find a counterexample?

42 AHSME Solutions

1. **(E)** If $a = 1$, $b = -2$ and $c = -3$, then
$$\boxed{a, b, c} = \frac{c+a}{c-b} = \frac{-3+1}{-3+2} = \frac{-2}{-1} = 2.$$

2. **(E)** Since $3 - \pi < 0$, we have $|3 - \pi| = -(3 - \pi) = \pi - 3$.

3. **(A)** $\left(4^{-1} - 3^{-1}\right)^{-1} = \left(\frac{1}{4} - \frac{1}{3}\right)^{-1} = \left(-\frac{1}{12}\right)^{-1} = -12.$

4. **(C)** A triangle cannot contain one $90°$ angle and another angle greater than $90°$ since the three angles must sum to $180°$. Triangles with sides having the following lengths show that each of the others is possible:
(A) $2, 3, 3$ **(B)** $1, 1, \sqrt{2}$ **(D)** $3, 4, 5$ **(E)** $3, 4, 6$

5. **(E)** Rectangle $CDEF$ has area $10 \cdot 20 = 200$. Triangle ABG is an isosceles right triangle. The altitude to base $\overline{BG}$ is $BG/2$, and $BG = BC + CF + FG = 5 + 10 + 5 = 20$, so the area of the triangle is $\frac{1}{2} \cdot 20 \cdot 10 = 100$. Thus the total area of the polygon is $200 + 100 = 300$.

OR

Extend $\overline{DC}$ to X and $\overline{EF}$ to Y where X and Y are on the line parallel to $\overline{DE}$ through A. By symmetry, $AX = AY = \frac{1}{2}DE = 5 = BC$. If $\overline{AB}$ intersects $\overline{DX}$ at W, then $\triangle AXW \cong \triangle BCW$ since $\angle WAX = \angle B = 45° = \angle BWC = \angle AWX$. Similarly, if $\overline{AG}$ intersects $\overline{EY}$ at

Z, then $\triangle AYZ \cong \triangle GFZ$. Hence the area of the arrow equals the area of rectangle $DXYE$ which is $10(20+5+5) = 300$.

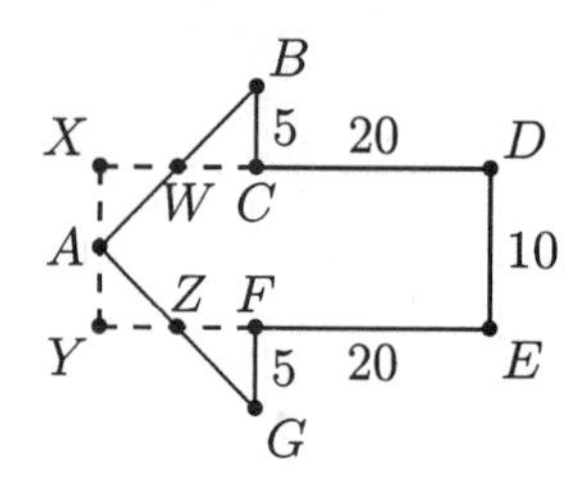

6. **(E)** $\sqrt{x\sqrt{x\sqrt{x}}} = \left(x\left(x \cdot x^{1/2}\right)^{1/2}\right)^{1/2} = \left(x\left(x^{3/2}\right)^{1/2}\right)^{1/2}$

$$= \left(x \cdot x^{3/4}\right)^{1/2} = \left(x^{7/4}\right)^{1/2} = x^{7/8} = \sqrt[8]{x^7}.$$

7. **(B)** Since $a = bx$,

$$\frac{a+b}{a-b} = \frac{bx+b}{bx-b} = \frac{b(x+1)}{b(x-1)} = \frac{x+1}{x-1}.$$

To show that the other four choices are incorrect, let $a = 2$ and $b = 1$, so $x = 2$ and

$$\frac{a+b}{a-b} = \frac{3}{1} \neq \begin{cases} \textbf{(A)}\ 2/(2+1) \\ \textbf{(C)}\ 1 \\ \textbf{(D)}\ 2-(1/2) \\ \textbf{(E)}\ 2+(1/2) \end{cases}$$

OR

Dividing numerator and denominator by b shows that

$$\frac{a+b}{a-b} = \frac{\frac{a}{b}+1}{\frac{a}{b}-1} = \frac{x+1}{x-1}.$$

Note. More generally, if $\dfrac{s}{t} = \dfrac{u}{v} \neq 1$ then

$$\frac{s+t}{s-t} = \frac{\frac{u}{v}t+t}{\frac{u}{v}t-t} = \frac{u+v}{u-v}.$$

8. **(C)** The volume of liquid X in cm^3 is $3 \cdot 6 \cdot 12 = 216$. The film is a cylinder whose volume in cm^3 is $0.1\pi r^2$. Solve $216 = 0.1\pi r^2$ to find $r = \sqrt{2160/\pi}$.

9. **(D)** Let the population be P at time $t = 0$, and suppose the population increased by $k\%$ from time $t = 0$ to time $t = 2$. The population at time $t = 2$

is
$$P \cdot \left(1 + \frac{k}{100}\right) = \left[P \cdot \left(1 + \frac{i}{100}\right)\right]\left(1 + \frac{j}{100}\right).$$

Hence
$$1 + \frac{k}{100} = 1 + \frac{i+j}{100} + \frac{ij}{10000},$$

so
$$k = i + j + \frac{ij}{100}.$$

Let $i = j = 10$ to show that the other choices are not always correct: If the population was $P = 100$ at $t = 0$ then it would be $P = 100 + 10 = 110$ at $t = 1$ and $P = 110 + 11 = 121$ at $t = 2$, an increase of 21%. Testing $i = j = 10$ in the answer choices, we have

(A) $(10+10)\% = 20\%$ **(B)** $(10)(10)\% = 100\%$
(C) $(10+100)\% = 110\%$ **(D)** $\left(10+10+\frac{100}{100}\right)\% = 21\%$
(E) $\left(10+10+\frac{20}{100}\right)\% = 20.2\%$

Comment. Due to the multiple choice nature of this examination, the above test with $i = j = 10$, or almost any other positive number of your choice, could be used to detect the correct choice.

10. **(B)** The longest chord through P is the diameter, $\overline{XY}$, which has length 30. The shortest chord through P, $\overline{CD}$, is perpendicular to this diameter. Hence its length is $2\sqrt{15^2 - 9^2} = 24$. As the chords rotate through point P, their lengths will take on all real numbers between 24 and 30 twice. [See figure.] Thus, for each of the five integers k strictly between 24 and 30 there are *two* chords of length k through P. This gives a total of $2 + 5 \cdot 2 = 12$ chords with integer lengths.

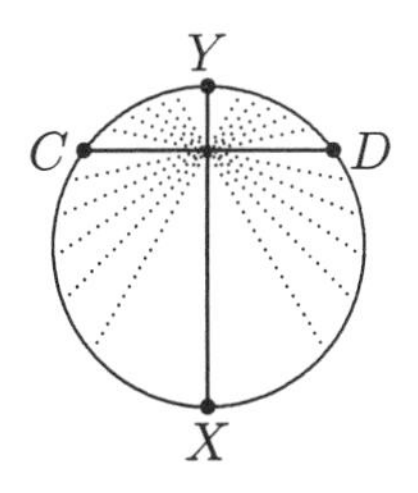

11. **(B)** Let x denote the distance in kilometers from the top of the hill to where they meet. When they meet, Jack has been running for $(5/15) + (x/20)$ hours and Jill has been running for $(5 - x)/16$ hours. Since Jack has been running $1/6$ hour longer than Jill, we solve

$$\left(\frac{5}{15} + \frac{x}{20}\right) - \frac{5-x}{16} = \frac{1}{6}$$

to find $x = 35/27$.

OR

Jack runs up the hill in 20 minutes. Therefore at the time when he starts down the hill, Jill has been running for 10 minutes and has come $16 \cdot (1/6) = 8/3$ km up the hill. Let t be the time needed to cover the $7/3$ km that now separates them. Then

$$20t + 16t = \frac{7}{3}, \quad \text{so} \quad t = \frac{7}{108}.$$

The distance from the top of the hill is the distance that Jack travels, namely $20 \cdot (7/108) = 35/27$ km.

OR

Sketch the distance, y, from the starting point versus the number of minutes, t, that Jack has been running. Since Jack runs up the hill in 20 minutes and back down in 15, Jack's graph consists of two line segments, one from $(0,0)$ to $(20,5)$ and the other from $(20,5)$ to $(35,0)$. The graph for Jill also consists of two line segments, the first of which is from $(10,0)$ with slope $16/60$ to $(28\frac{3}{4}, 5)$.

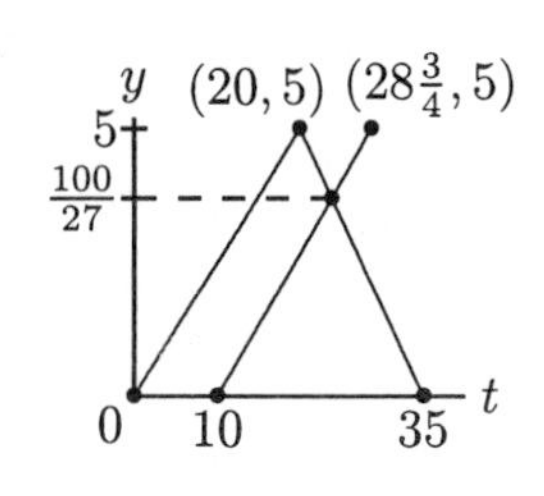

The y-coordinate of the intersection of this segment with Jack's second segment is the distance from the starting point where the joggers meet. We solve

$$y - 5 = \frac{0-5}{35-20}(t - 20)$$

and

$$y = \frac{16}{60}(t - 10)$$

simultaneously to find that $y = 100/27$. Hence the joggers meet $5 - (100/27) = 35/27$ km from the top of the hill.

12. **(D)** Let d be the common difference of the arithmetic sequence. The sum of the degrees in the interior angles of this hexagon is

$$m + (m-d) + (m-2d) + \cdots + (m-5d) = 6m - 15d.$$

The sum of the interior angles of any hexagon is $(6-2)180^\circ = 720^\circ$. Hence,

$$6m - 15d = 720 \qquad \text{or} \qquad 6m = 15d + 720 = 5(3d + 144),$$

so m is divisible by 5. Because the hexagon is convex, it follows that $m < 180$, and hence $m \le 175$. Since

$$65 + 87 + 109 + 131 + 153 + 175 = 720,$$

there is such a hexagon and $m^\circ = 175^\circ$.

OR

Let a be the smallest angle and d be the common difference of the arithmetic sequence. Then, the sum of the degrees in all six interior angles is

$$a + (a+d) + (a+2d) + \cdots + (a+5d) = 6a + 15d,$$

so $6a + 15d = 720$ and thus $2a + 5d = 240$. But the sum of the smallest and largest angles in the hexagon is

$$a + (a + 5d) = 2a + 5d = 240.$$

Since $2a = 240 - 5d = 5(48 - d)$, a must be divisible by 5, and hence the smallest and largest angles in the hexagon are divisible by 5. Therefore, the largest candidate for an interior angle is 175°, and we verify that there is a hexagon with this interior angle as in the previous solution.

OR

Derive $2a + 5d = 240$ as above. Then note that the solutions to this equation are all of the form $(a, d) = (120-5k, 2k)$ for k some integer. Since the hexagon is convex, the largest angle,

$$a + 5d = (2a+5d) - a = 240 - (120 - 5k) = 120 + 5k < 180.$$

The largest k that satisfies this inequality is $k = 11$, and it produces $a + 5d = 120 + 55 = 175$ as a candidate for the largest interior angle. Verify this candidate as in the first solution.

13. **(D)** The probability that X wins is $1/(3+1)$ and the probability that Y wins is $3/(2+3)$. The sum of the winning probabilities for all

three horses must be 1, so the probability that Z wins is

$$1-\frac{1}{4}-\frac{3}{5}=\frac{3}{20}=\frac{3}{17+3}.$$

Hence the odds against Z winning are 17-to-3.

14. **(C)** The cubes

$$x=1,2^3,2^6,\ldots,2^{3k},\ldots,(2^{67})^3$$

have

$$d=1,4,7,\ldots,3k+1,\ldots,202$$

divisors, respectively. In fact, for any prime p, $(p^{67})^3$ has 202 divisors. To show that, of the choices listed, $d=202$ is the only possible answer, we prove that for *any* perfect cube $x>1$, d *must* be of the form $3k+1$:

If the cube x is written as $x=p_1^{3b_1}p_2^{3b_2}\cdots p_n^{3b_n}$ where the p_i are distinct primes, then its divisors are all the numbers of the form $p_1^{a_1}p_2^{a_2}\cdots p_n^{a_n}$, with $0\le a_i\le 3b_i$ for $i=1,2,\ldots,n$. Taking the product of the number of choices for each a_i yields $d=(3b_1+1)(3b_2+1)\cdots(3b_n+1)=3k+1$ for some integer k.

15. **(B)** Divide the chairs around the table into $60/3=20$ sets of three consecutive seats. If fewer than 20 people are seated at the table, then at least one of these sets of three seats will be unoccupied. If the next person sits in the center of this unoccupied set, then that person will not be seated next to anyone already seated. On the other hand, if 20 people are already seated, and each occupies the center seat in one of the sets of three, then the next person to be seated must sit next to one of these 20 people.

OR

If the next person who sits must sit next to an occupied seat, then there can be at most two empty chairs between any two (consecutive) occupied chairs. An arrangement with every third seat occupied is possible since the number of chairs is a multiple of 3, and thus the required minimum is $60/3=20$.

OR

Test, beginning with smaller numbers of chairs, n. If $n = 1, 2, 3$ then the minimum is 1; if $n = 4, 5, 6$ then the minimum is 2; if $n = 7, 8, 9$ then the minimum is 3. The pattern will continue so that if $n = 58, 59, 60$ then the minimum is 20.

16. **(D)** If s seniors took the **AHSME** then $3s/2$ non-seniors took it, so $s + (3s/2) = 100$, $s = 40$ and $3s/2 = 60$. If a is the mean score for the seniors then $(2/3)a$ is the mean score for the non-seniors, so

$$\frac{40a + 60\left(\frac{2}{3}a\right)}{100} = 100$$

and $a = 125$.

OR

If s is the number of seniors and a is their average score, then consider the weighted average:

$$100 = \frac{sa + \left(\frac{3}{2}s\right)\left(\frac{2}{3}a\right)}{s + \frac{3}{2}s} = \frac{2a}{5/2} = \frac{4}{5}a.$$

Solve to find that $a = 125$.

OR

If s is the number of seniors who took the **AHSME** then $3s/2$ is the number of non-seniors. Since $s + (3s/2) = 100$, the number of seniors is $s = 40$ and the number of non-seniors is $3s/2 = 60$. Let $t_1, t_2, \ldots, t_{40}$ be the scores for the seniors, and $t_{41}, t_{42}, \ldots, t_{100}$ be the scores for the non-seniors. Then the sum of all the scores is

$$\sum_{i=1}^{100} t_i = 100 \cdot 100 = 10000.$$

Let the sum of the seniors' scores be

$$M = \sum_{i=1}^{40} t_i.$$

Then the sum of the scores of the non-seniors will be

$$\sum_{i=41}^{100} t_i = 10000 - M.$$

Since the seniors' average score was $3/2$ of the non-seniors' average,

$$\frac{M}{40} = \frac{10000 - M}{60} \cdot \frac{3}{2},$$

whose solution is $M = 5000$. Thus the mean score for the seniors was

$$\frac{M}{40} = \frac{5000}{40} = 125$$

and the mean score for non-seniors was

$$\frac{10000 - M}{60} = \frac{5000}{60} = 83\frac{1}{3}.$$

Note. The answer is independent of the number of students taking the test, so long as that number is divisible by 5 since there must be an integer solution to $s + (3s/2) = T$ where T is the total number of students.

17. **(D)** Since a palindrome between 1000 and 2000 begins and ends with a 1, there are 10 numbers, all of the form $1dd1$ to check. Since 3 divides 1221, 1551, and 1881, 7 divides 1001 and 1771, and $1331 = 11^3$, these six choices can be eliminated. We then note that the remaining four numbers, $1111 = 11 \cdot 101$, $1441 = 11 \cdot 131$, $1661 = 11 \cdot 151$ and $1991 = 11 \cdot 181$ all have both required properties.

OR

First note that 11 is the only two-digit prime palindrome. The three-digit palindrome must be less than $2000/11 = 181.\overline{81}$. The only three-digit prime palindromes in the range 100 to $2000/11$ are 101, 131, 151 and 181. Thus, $1111 = 11 \cdot 101$, $1441 = 11 \cdot 131$, $1661 = 11 \cdot 151$ and $1991 = 11 \cdot 181$ are the only four numbers with the two required properties.

OR

A palindrome between 1000 and 2000 has the form $1000 + 100b + 10b + 1 = 11(10b + 91)$. Since 11 is the only 2-digit prime palindrome, we investigate to discover when $10b + 91$ is a 3-digit prime palindrome. It must have the form $100 + 10c + 1$ and not exceed $2000/11$, so $0 \le c \le 8$. With c in this range, $100 + 10c + 1$ will be prime for $c = 0, 3, 5, 8$. The corresponding years are 1111, 1441, 1661, and 1991.

18. **(D)** The set S consists of all complex numbers of the form

$$z = \frac{r}{3 + 4i}\left(\frac{3 - 4i}{3 - 4i}\right) = \frac{r}{25}(3 - 4i)$$

for some real number r. Since S consists of all real multiples of $3-4i$, each point in S is on the line through the origin and $3-4i$, and conversely.

OR

Let $z = x + iy$. Then

$$(3+4i)z = (3+4i)(x+iy) = 3x - 4y + (3y+4x)i,$$

which is real if and only if $3y + 4x = 0$, the equation of a line.

OR

Let $z = r(\cos\theta + i\sin\theta)$ and $3 + 4i = 5(\cos\alpha + i\sin\alpha)$, where $\cos\alpha = 3/5$ and $\sin\alpha = 4/5$. Then

$$(3+4i)z = 5r(\cos(\alpha+\theta) + i\sin(\alpha+\theta))$$

is real if and only if $\theta = -\alpha + n\pi$ for some integer n. For fixed n, the set of all z of the form

$$r(\cos(-\alpha+n\pi) + i\sin(-\alpha+n\pi))$$

for some real r is the set of points on the line in the second and fourth quadrants which forms an angle of $|\alpha|$ at the origin with the x-axis.

19. **(B)** Since $AB = \sqrt{3^2+4^2} = 5$ and $BD = \sqrt{5^2+12^2} = 13$, it follows that

$$\begin{aligned}\frac{m}{n} = \frac{DE}{DB} &= \sin\angle DBE = \sin(180^\circ - \angle DBE)\\ &= \sin\angle DBC = \sin(\angle DBA + \angle ABC)\\ &= \sin(\angle DBA)\cos(\angle ABC) + \cos(\angle DBA)\sin(\angle ABC)\\ &= \frac{12}{13}\cdot\frac{4}{5} + \frac{5}{13}\cdot\frac{3}{5} = \frac{63}{65}\end{aligned}$$

and $m + n = 128$.

OR

Draw $\overline{AG}$ parallel to $\overline{CE}$ with G on $\overline{DE}$. Then $\angle GAD = \angle CAB$ since both are complementary to $\angle GAB$. Thus, $\triangle GAD \sim \triangle CAB$, and

$$\frac{DG}{BC} = \frac{AD}{AB} = \frac{12}{\sqrt{3^2+4^2}}.$$

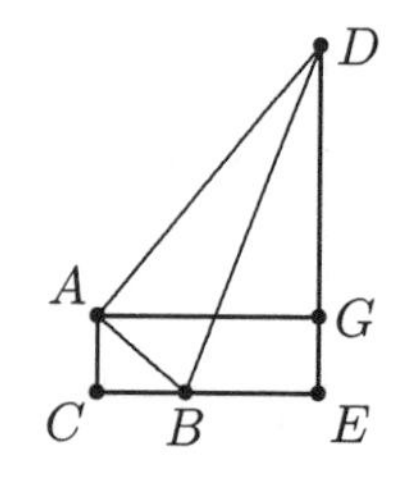

Hence $DG = (12/5)BC = 48/5$, so

$$DE = DG + GE = DG + AC$$
$$= \frac{48}{5} + 3 = \frac{63}{5}.$$

Therefore,

$$\frac{DE}{DB} = \frac{63/5}{\sqrt{12^2+5^2}} = \frac{63}{65},$$

so $m + n = 128$.

OR

Extend $\overline{DA}$ and $\overline{BC}$ to meet at F. Since $\angle BAC$ and $\angle AFC$ are both complements to $\angle FAC$, $\triangle AFC \sim \triangle BAC$. Thus

$$\frac{AF}{BA} = \frac{AC}{BC} = \frac{FC}{AC}, \text{ or } \frac{AF}{5} = \frac{3}{4} = \frac{FC}{3},$$

$$\text{so } AF = \frac{15}{4} \text{ and } FC = \frac{9}{4}.$$

Since $\triangle AFC \sim \triangle DFE$, it follows that

$$\frac{DE}{AC} = \frac{DF}{AF} \text{ or } \frac{DE}{3} = \frac{\frac{15}{4}+12}{\frac{15}{4}}.$$

Thus $DE = 63/5$. Therefore,

$$\frac{DE}{DB} = \frac{63/5}{13} = \frac{63}{65},$$

so $m + n = 128$.

Note. Observe that trapezoid $ACED$ is partitioned into three triangles by $\overline{AB}$ and $\overline{BD}$, and that the area of the trapezoid and the sum of the areas of the three triangles can be expressed in terms of DE:

$$[ACED] = \frac{1}{2}(3+DE)\left(4+\sqrt{169-DE^2}\right);$$

$$[ACB] + [ABD] + [BDE] = 6 + 30 + \frac{1}{2}DE\sqrt{169-DE^2}.$$

One can equate these areas, subtract $\frac{1}{2}DE\sqrt{169 - DE^2}$ from both sides, square to eliminate the radical, factor the resulting quadratic equation, and argue away the extraneous root for DE. However, this method is so much more computationally intensive than the above methods that it is not recommended.

20. **(E)** By the Binomial Theorem, for any a and b

$$(a+b)^3 = a^3 + 3a^2b + 3ab^2 + b^3.$$

For $a = 2^x - 4$ and $b = 4^x - 2$, we have

$$a + b = 4^x + 2^x - 6,$$

and therefore we are given

$$a^3 + b^3 = (a+b)^3.$$

Hence $3a^2b + 3ab^2 = 0$. But,

$$3ab(a+b) = 0 \iff a = 0,\ b = 0,\ \text{or } a + b = 0.$$

Thus

$$\begin{aligned} a = 2^x - 4 &= 0; \quad \text{i.e., } x = 2; \\ \text{or } b = 4^x - 2 &= 0; \quad \text{i.e., } x = \frac{1}{2}; \\ \text{or } a + b = 4^x + 2^x - 6 = (2^x + 3)(2^x - 2) &= 0; \quad \text{i.e., } x = 1. \end{aligned}$$

Note that $2^x + 3 = 0$ has no real roots. Therefore, the sum of the real roots is

$$2 + \frac{1}{2} + 1 = \frac{7}{2}.$$

Note. Many times equations involving 2^x and 4^x can be simplified and solved with the substitution $y = 2^x$ since $4^x = 2^{2x} = (2^x)^2 = y^2$. That method was used implicitly near the end in solving $4^x + 2^x - 6 = 0$, but that method is not efficient as a first step in this particular problem.

21. **(A)** If $\dfrac{x}{x-1} = \sec^2\theta$ then

$$x = x\sec^2\theta - \sec^2\theta$$
$$x(\sec^2\theta - 1) = \sec^2\theta$$
$$x\tan^2\theta = \sec^2\theta.$$

Hence

$$x = \frac{\sec^2\theta}{\tan^2\theta} = \frac{1}{\sin^2\theta}$$

and $f(\sec^2\theta) = \sin^2\theta$.

OR

First solve $y = \dfrac{x}{x-1}$ for x to find $x = \dfrac{y}{y-1}$. Then $f(y) = \dfrac{y-1}{y}$. Hence

$$f(\sec^2\theta) = \frac{\sec^2\theta - 1}{\sec^2\theta} = 1 - \cos^2\theta = \sin^2\theta.$$

OR

Since $\dfrac{1}{1-\frac{1}{x}} = \dfrac{x}{x-1}$, $f\left(\dfrac{1}{1-t}\right) = t$. Thus

$$f\left(\sec^2\theta\right) = f\left(\frac{1}{\cos^2\theta}\right) = f\left(\frac{1}{1-\sin^2\theta}\right) = \sin^2\theta.$$

22. **(B)** Let C be the center of the smaller circle, T be the point where the two circles are tangent, and X be the intersection of the common internal tangent with $\overline{AB}$. Since tangents from a common point are equal, $BX = TX = AX = AB/2 = 2$. Since $\triangle ACP \sim \triangle TXP$, it follows that

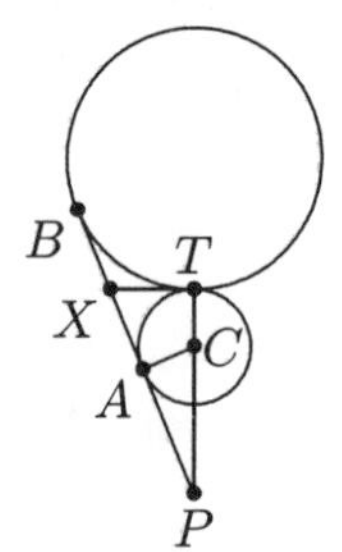

$$\frac{AC}{TX} = \frac{AP}{TP}, \quad \text{or} \quad \frac{AC}{2} = \frac{4}{\sqrt{6^2 - 2^2}},$$

so $AC = \sqrt{2}$. Hence the area of the circle with radius AC is 2π.

OR

Let C_1, C_2, r and R be the centers and radii of the smaller and larger circles, respectively. Points P, C_1 and C_2 are collinear by symmetry. Since the right triangles PAC_1 and PBC_2 are similar,

$$\frac{r}{R} = \frac{PC_1}{PC_2} = \frac{PA}{PB} = \frac{4}{8}.$$

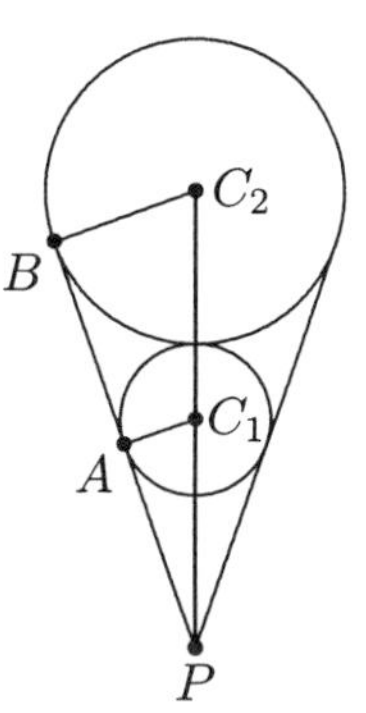

Thus $R = 2r$ and

$$PC_1 = C_1C_2 = R + r = 3r.$$

Apply the Pythagorean theorem to $\triangle PAC_1$ to find $4^2 + r^2 = (3r)^2$, $r^2 = 2$ and $\pi r^2 = 2\pi$.

23. **(C)** Use coordinates with $B = (0,0)$, $F = (1,0)$ and $E = (0,1)$. The equations of lines $\overline{BH}$, $\overline{IH}$ and $\overline{EI}$ are

$$y = x,$$
$$y = -2x + 2$$
$$\text{and} \quad y = \frac{1}{2}x + 1,$$

respectively. Solve pairs of these equations simultaneously to find that

$$H = \left(\frac{2}{3}, \frac{2}{3}\right) \quad \text{and} \quad I = \left(\frac{2}{5}, \frac{6}{5}\right).$$

The altitude of $\triangle BHF$ from vertex H is the y-coordinate of H, $2/3$. Since $BF = 1$, it follows that $[BHF] = 1/3$. Similarly, the altitude of $\triangle AIE$ from vertex I is the x-coordinate of I, $2/5$, and $AE = 1$, so $[AIE] = 1/5$. Since $AB = 2$ is the length of the altitude to base $\overline{BF}$ in $\triangle BAF$, $[BAF] = 1$. Thus

$$[BEIH] = [BAF] - [AIE] - [BHF] = 1 - \frac{1}{5} - \frac{1}{3} = \frac{7}{15}.$$

OR

Compute $[AIE] = 1/5$ as above and note that the altitude of $\triangle AHB$ from H is $2/3$, so $[AHB] = 2/3$ and

$$[BEIH] = [AHB] - [AIE] = \frac{2}{3} - \frac{1}{5} = \frac{7}{15}.$$

OR

Triangles DAE and ABF have equal sides so they are congruent, and thus $\angle AEI = \angle BFA$. Hence $\triangle AIE$ is a right triangle similar to $\triangle ABF$. Since $[ABF] = 1$ and $AF = \sqrt{1^2 + 2^2} = \sqrt{5}$, we have

$$[AIE] = \frac{[AIE]}{1} = \frac{[AIE]}{[ABF]} = \frac{AE^2}{AF^2} = \frac{1}{5}.$$

Note that $\triangle BHF$ is similar to $\triangle DHA$ because of equal angles, and that the ratio of similarity is $BF/DA = 1/2$. Hence

$$\frac{HF}{HA} = \frac{1}{2} \quad \text{so} \quad \frac{HF}{AF} = \frac{1}{3}.$$

Thus, since $\triangle BHF$ and $\triangle BAF$ share side $\overline{BF}$ and the altitudes to that side are in the ratio $1 : 3$,

$$[BHF] = \frac{[BHF]}{1} = \frac{[BHF]}{[BAF]} = \frac{1}{3}.$$

Again $[BEIH] = [BAF] - [AIE] - [BHF] = 1 - (1/5) - (1/3) = 7/15$.

OR

Let the areas of triangles AEI, EHI, BHE and BHF be w, x, y and z, respectively. Since $BE = BF$ and $\angle EBH = \angle FBH$, triangles BHE and BHF are congruent. Hence, $y = z$. Since $\overline{EH}$ is a median of $\triangle AHB$, we have $w + x = y$. Therefore

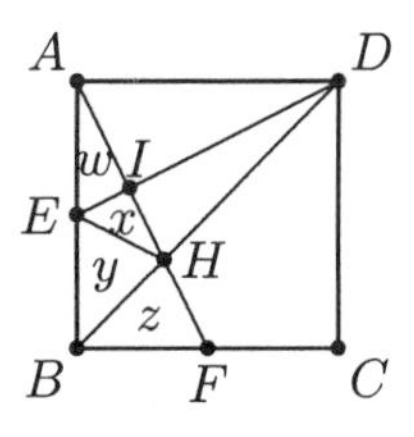

$$3y = (w + x) + y + z = \frac{1}{2} BF \cdot AB = 1, \quad \text{or} \quad y = \frac{1}{3}.$$

If $\triangle ABF$ is rotated 90° clockwise about the center of the square, it coincides with $\triangle DAE$. Hence $\overline{AF} \perp \overline{DE}$, from which it follows that $\triangle AIE \sim \triangle DAE$. Since $AE/DE = 1/\sqrt{5}$,

$$w = [AIE] = \frac{[AIE]}{1} = \frac{[AIE]}{[DAE]} = \left(\frac{1}{\sqrt{5}}\right)^2 = \frac{1}{5}.$$

Since $w + x = y$, we have

$$[BEIH] = x + y = (w + x) + y - w = 2y - w = \frac{2}{3} - \frac{1}{5} = \frac{7}{15}.$$

OR

The three lines, $\overline{BD}$ and the lines through E and F parallel to $\overline{BD}$, are equally spaced, so $HF = AF/3$ and hence, comparing their altitudes to $\overline{BF}$,

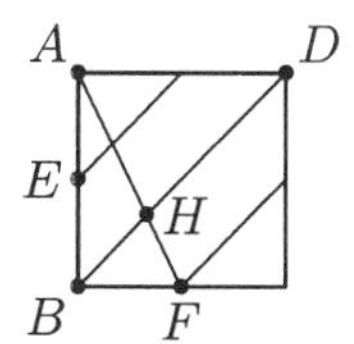

$$[HBF] = \frac{1}{3}[ABF] = \frac{1}{3}.$$

Next choose X and Y on $\overline{AB}$ extended so $\overline{ED} \parallel \overline{BM} \parallel \overline{XN} \parallel \overline{YC}$ with $\overline{XN}$ through F. Since F is the midpoint of $\overline{BC}$ and thus X is the midpoint of $\overline{BY}$, we have

$$AE : EB : BX : XY :: 2 : 2 : 1 : 1,$$

so

$$\frac{AI}{AF} = \frac{AE}{AX} = \frac{2}{5}$$

and $[ABI] = (2/5)[ABF]$. (Compare their altitudes to $\overline{AB}$.) Since E is the midpoint of $\overline{AB}$,

$$[AEI] = \frac{1}{2}[ABI] = \frac{1}{2}\left(\frac{2}{5}[ABF]\right) = \frac{1}{5}(1) = \frac{1}{5}.$$

Again $[BEIH] = [ABF] - [AEI] - [HBF] = 1 - (1/3) - (1/5) = 7/15$.

OR

Coordinatize the problem as in the first solution. In particular, note that

$$B = (0,0), \quad E = (0,1), \quad I = \left(\frac{2}{5}, \frac{6}{5}\right), \quad H = \left(\frac{2}{3}, \frac{2}{3}\right).$$

If the coordinates of a triangle counterclockwise around the triangle are

$$(x_1, y_1), \quad (x_2, y_2), \quad (x_3, y_3),$$

then the area of the triangle is half the value of the determinant

$$\begin{vmatrix} 1 & x_1 & y_1 \\ 1 & x_2 & y_2 \\ 1 & x_3 & y_3 \end{vmatrix}.$$

Hence

$$\begin{aligned}
[BEIH] &= [BHI] + [BIE] \\
&= \frac{1}{2}\begin{vmatrix} 1 & 0 & 0 \\ 1 & 2/3 & 2/3 \\ 1 & 2/5 & 6/5 \end{vmatrix} + \frac{1}{2}\begin{vmatrix} 1 & 0 & 0 \\ 1 & 2/5 & 6/5 \\ 1 & 0 & 1 \end{vmatrix} \\
&= \frac{1}{2}\left(\frac{8}{15}\right) + \frac{1}{2}\left(\frac{2}{5}\right) = \frac{7}{15}.
\end{aligned}$$

24. **(D)** The 90° rotation relates each (x, y) on G' to the point $(x \cos 90^\circ + y \sin 90^\circ, -x \sin 90^\circ + y \cos 90^\circ)$ on G. Thus, the point (x, y) is on the graph of G' if and only if the point $(y, -x)$ is on the graph of G, so $-x = \log_{10} y$. This last equation is equivalent to $y = 10^{-x}$, which is an equation for G'. Since $(x, y) = (10, 1)$ is on G, it follows that $(x, y) = (-1, 10)$ must be on G', which shows that no other choice is correct.

Note. The answer can also be found by elimination in two ways: (*i*) Point $(x, y) = (10, 1)$ is on G. This point rotates to $(x, y) = (-1, 10)$ which must be on G'. Only choice **(D)** contains this point.

(*ii*) Eliminate all but the correct answer by first quadrant sketches: Sketch part of G in the fourth quadrant, rotate it 90° counterclockwise to get a sketch of part of G' in the first quadrant, and then sketch a part of choices **(A)** through **(E)** in the first quadrant to note that **(D)** must be the answer.

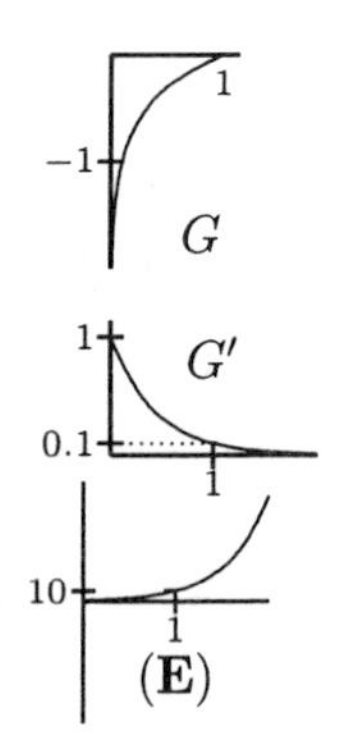

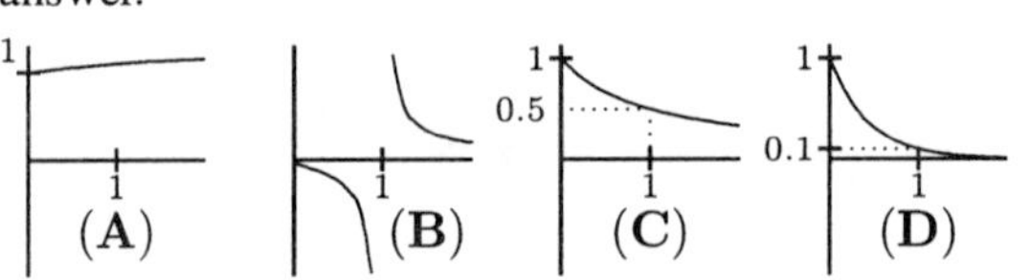

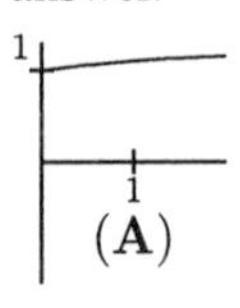

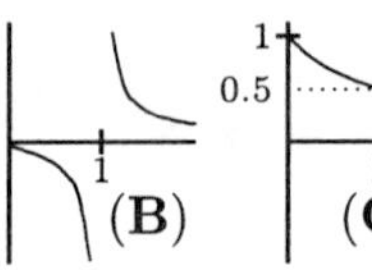

25. **(D)** Since $T_n = n(n+1)/2$,

$$\begin{aligned}
P_n &= \frac{T_n}{T_n - 1} P_{n-1} \\
&= \frac{n(n+1)/2}{[n(n+1) - 2]/2} P_{n-1} = \frac{(n+1)n}{(n+2)(n-1)} P_{n-1}.
\end{aligned}$$

Therefore

$$\begin{aligned}
P_{1991} &= \frac{1992 \cdot 1991}{1993 \cdot 1990} \cdot \left(\frac{1991 \cdot 1990}{1992 \cdot 1989} P_{1989} \right) \\
&= \frac{1991}{1993} \cdot \frac{1991}{1989} P_{1989} \\
&= \frac{1991}{1993} \cdot \frac{1991}{1989} \cdot \left(\frac{1990 \cdot 1989}{1991 \cdot 1988} P_{1988} \right) \\
&= \frac{1991}{1993} \cdot \frac{1990}{1988} P_{1988} \\
&= \cdots = \frac{1991}{1993} \cdot \frac{k+2}{k} P_k \\
&= \cdots = \frac{1991}{1993} \cdot \frac{4}{2} P_2 = \frac{1991}{1993} \cdot 3,
\end{aligned}$$

so 2.9 is closest to P_{1991}.

OR

Since

$$P_n = \prod_{k=2}^{n} \frac{k(k+1)}{(k+2)(k-1)} = \frac{\left(\prod_{k=2}^{n} k\right)\left(\prod_{k=2}^{n} (k+1)\right)}{\left(\prod_{k=2}^{n} (k+2)\right)\left(\prod_{k=2}^{n} (k-1)\right)}$$

$$= \frac{n! \left(\frac{(n+1)!}{2}\right)}{\left(\frac{(n+2)!}{2 \cdot 3}\right)(n-1)!} = \frac{3n}{n+2},$$

so $P_{1991} = (3 \cdot 1991)/1993$ which is closest to 2.9.†

26. **(C)** Let $abcdef$ be a six-digit cute integer. Since the five-digit number $abcde$ is divisible by 5, the fifth digit, e, must be 5. Since ab is divisible by 2, $abcd$ is divisible by 4, and $abcdef$ is divisible by 6, it follows that the second, fourth and sixth digits, b, d and f must be even. Since no digit can be larger than 6,

$$\{b, d, f\} = \{2, 4, 6\} \quad \text{and} \quad \{a, c\} = \{1, 3\}.$$

† **Notation:** $\prod_{k=m}^{n} a_k = a_m \cdot a_{m+1} \cdot a_{m+2} \cdot \cdots \cdot a_n.$

Since a and c must be 1 and 3 in some order, and b must be 2 or 4 or 6, it follows that

$$a+b+c=\begin{cases} a+2+c=4+2=6 & \text{or} \\ a+4+c=4+4=8 & \text{or} \\ a+6+c=4+6=10. \end{cases}$$

Since abc is divisible by 3, $a+b+c$ is divisible by 3 and the second digit must be $b=2$. With $d=4$ or $d=6$, the four-digit number $123d$ or $321d$ must be divisible by 4. Neither 1234 nor 3214 is divisible by 4, so the fourth digit must be $d=6$, leaving and $f=4$ for the sixth digit. Therefore there are two cute 6-digit integers, 123654 and 321654.

27. **(C)** Clear the denominator in the first equation to obtain

$$\begin{aligned} \left(x^2-(x^2-1)\right)+1 &= 20\left(x-\sqrt{x^2-1}\right) \\ (1)+1 &= 20\left(x-\sqrt{x^2-1}\right) \\ x-\sqrt{x^2-1} &= \frac{1}{10}. \end{aligned}$$

Observe that since the first term above simplified to (1),

$$\left(x-\sqrt{x^2-1}\right)\left(x+\sqrt{x^2-1}\right)=1. \qquad (*)$$

Therefore

$$x+\sqrt{x^2-1}=\frac{1}{x-\sqrt{x^2-1}}=10.$$

Consequently,

$$\begin{aligned} 2x &= \left(x-\sqrt{x^2-1}\right)+\left(x+\sqrt{x^2-1}\right) \\ &= \frac{1}{1/(x-\sqrt{x^2-1})}+\left(x+\sqrt{x^2-1}\right) \\ &= \frac{1}{10}+10=10.1. \end{aligned}$$

Substitute x^2 for x in $(*)$ to show that

$$\left(x^2-\sqrt{x^4-1}\right)\left(x^2+\sqrt{x^4-1}\right)=1.$$

Thus

$$\begin{aligned} x^2+\sqrt{x^4-1}&+\frac{1}{x^2+\sqrt{x^4-1}} \\ &= x^2+\sqrt{x^4-1}+\left(x^2-\sqrt{x^4-1}\right) \\ &= 2x^2 = 2\left(\frac{10.1}{2}\right)^2 = 51.005. \end{aligned}$$

28. **(B)** Since the number of white marbles is either unchanged or decreases by 2 after each replacement, the number of white marbles remains even. Since every set removed that includes at least one white marble is replaced by a set containing at least one white marble, the number of white marbles can never be zero. Note that **(B)** is the only choice including at least two white marbles. We can attain this result in many ways. One way is to remove 3 white marbles 49 times to arrive at 149 black and 2 white marbles, and then remove 1 black and 2 white marbles 149 times.

    Note. The concept of a *loop invariant* is an important concept in computer science. The fact that the number of white marbles must always be even is an example of a loop invariant in this procedure.

29. **(B)** Sum the angles in $\triangle A'PB$, and then sum the angles that make up the straight angle at A':

$$\begin{aligned} \angle BA'P + \angle A'PB + 60^\circ &= 180^\circ \\ \angle BA'P + 60^\circ + \angle QA'C &= 180^\circ. \end{aligned}$$

It follows that $\angle A'PB = \angle QA'C$ and thus $\triangle A'PB \sim \triangle QA'C$. Let $x = AP = A'P$ and $y = QA = QA'$. Then

$$\frac{A'P}{QA'} = \frac{A'B}{QC} = \frac{PB}{A'C}, \quad \text{or} \quad \frac{x}{y} = \frac{1}{3-y} = \frac{3-x}{2}.$$

Equate the first fraction to the second fraction, and then the first to the third and simplify:

$$\begin{aligned} 3x - xy &= y \\ 3y - xy &= 2x. \end{aligned}$$

Eliminate the xy term to find that $5x = 4y$. Substitute this into the equations to obtain $x = 7/5$ and $y = 7/4$. Now apply the Law of Cosines to $\triangle PAQ$,

$$PQ^2 = x^2 + y^2 - 2xy\cos 60^\circ = \frac{49}{25} + \frac{49}{16} - \frac{49}{20} = \frac{49 \cdot 21}{400},$$

which leads to $PQ = (7/20)\sqrt{21}$.

OR

Let $x = PA = PA'$ and $y = QA = QA'$. Apply the Law of Cosines to $\triangle PBA'$ to obtain

$$x^2 = (3-x)^2 + 1^2 - 2(3-x)\cos 60^\circ$$

which leads to $x = 7/5$. Consider $\triangle QCA'$ in a similar fashion to find $y = 7/4$. Then complete the solution as above.

30. **(B)** If a set has k elements, then it has 2^k subsets. Thus we are given

$$2^{100} + 2^{100} + 2^{|C|} = 2^{|A \cup B \cup C|}$$
$$2^{101} + 2^{|C|} = 2^{|A \cup B \cup C|}$$
$$1 + 2^{|C|-101} = 2^{|A \cup B \cup C|-101}.$$

The left side, $1 + 2^{|C|-101}$, is larger than 1, so the right side is also larger than $1 = 2^0$. Therefore $|A \cup B \cup C| - 101 > 0$, and it is an integer. Thus the left side, $1 + 2^{|C|-101}$ must be an integral power of 2. The only integral power of 2 of the form $1 + 2^m$ is $1 + 2^0 = 2^1$. Hence

$$|C| = 101 \quad \text{and} \quad |A \cup B \cup C| = 102.$$

The inclusion-exclusion formula for three sets is $|A \cup B \cup C|$

$$= |A| + |B| + |C| - |A \cap B| - |A \cap C| - |B \cap C| + |A \cap B \cap C|.$$

Since $|A \cup B \cup C| = 102$, $|A| = |B| = 100$ and $|C| = 101$, we have

$$102 = 301 - |A \cap B| - |A \cap C| - |B \cap C| + |A \cap B \cap C|.$$

Solve this equation for $|A \cap B \cap C|$:

$$|A \cap B \cap C| = |A \cap B| + |A \cap C| + |B \cap C| - 199.$$

Now use $|X \cap Y| = |X|+|Y|-|X \cup Y|$ (obtained from the inclusion-exclusion formula for any two sets X and Y):

$$\begin{aligned}|A\cap B\cap C| &= (|A|+|B|-|A\cup B|)+(|A|+|C|-|A\cup C|)\\ &\quad +(|B|+|C|-|B\cup C|)-199\\ &= (2\,|A|+2\,|B|+2\,|C|-199)-(|A\cup B|\\ &\quad +|A\cup C|+|B\cup C|)\\ &= 403-(|A\cup B|-|A\cup C|-|B\cup C|)\,.\end{aligned}$$

Since

$$A\cup B,\ A\cup C,\ B\cup C\subseteq A\cup B\cup C,$$

we have

$$|A\cup B|\,,\ |A\cup C|\,,\ |B\cup C|\le 102.$$

Thus

$$\begin{aligned}|A\cap B\cap C| &= 403-(|A\cup B|-|A\cup C|-|B\cup C|)\\ &\ge 403-3\cdot 102=97.\end{aligned}$$

The example

$$A=\{1,2,\ldots,100\},\qquad B=\{3,4,\ldots,102\},$$
$$C=\{1,2,4,5,6,\ldots,101,102\}$$

shows that

$$|A\cap B\cap C|=|\{4,5,6,\ldots,100\}|=97$$

is possible.

43 AHSME Solutions

1. **(B)** $6(8x + 10\pi) = [2(3)][2(4x + 5\pi)] = (2 \cdot 2)[3(4x + 5\pi)] = 4P.$

OR

Since $P = 12x + 15\pi$, we have

$$6(8x + 10\pi) = 48x + 60\pi = 4(12x + 15\pi) = 4P.$$

OR

Since $3(4x + 5\pi) = P$, it follows that

$$x = \frac{1}{4}\left(\frac{P}{3} - 5\pi\right).$$

Hence

$$\begin{aligned} 6[8x + 10\pi] &= 6\left[8\left(\frac{1}{4}\right)\left(\frac{P}{3} - 5\pi\right) + 10\pi\right] \\ &= 6\left[\left(\frac{2}{3}P - 10\pi\right) + 10\pi\right] = 6\left[\frac{2}{3}P\right] = 4P. \end{aligned}$$

2. **(B)** Of the 80% which are coins, 60% are gold. This is 60% of 80%, or 48% of the objects in the urn.

3. **(C)** Since the slope of the line through points (x_1, y_1) and (x_2, y_2) is $(y_2 - y_1)/(x_2 - x_1)$, it follows that

$$m = \frac{m-3}{1-m}.$$

Therefore,

$$m - m^2 = m - 3; \quad \text{i.e., } m^2 = 3.$$

Since $m > 0$, it follows that $m = \sqrt{3}$.

4. **(A)** The number 3^a is odd, and $(b-1)$ is even. Therefore $(b-1)^2$ is even and so is $(b-1)^2 c$. Hence the sum

$$3^a + (b-1)^2 c$$

is odd for all choices of c.

Note. The parity of the expression is independent of the parity of a since 3^a is odd whether a is even or odd.

5. **(B)** $6^6 + 6^6 + 6^6 + 6^6 + 6^6 + 6^6 = 6(6^6) = 6^1 \cdot 6^6 = 6^{1+6} = 6^7$.

6. **(D)** $\dfrac{x^y y^x}{y^y x^x} = \dfrac{x^{y-x}}{y^{y-x}} = \left(\dfrac{x}{y}\right)^{y-x}$ One may verify that none of the other choices is correct by substituting $x = 2$ and $y = 1$.

OR

$$\frac{x^y y^x}{y^y x^x} = \left(\frac{x^y}{x^x}\right)\left(\frac{y^x}{y^y}\right) = x^{y-x} y^{x-y}$$

$$= x^{y-x} y^{-(y-x)} = \left(\frac{x}{y}\right)^{y-x}.$$

7. **(B)** $\dfrac{w}{y} = \dfrac{w}{x} \cdot \dfrac{x}{z} \cdot \dfrac{z}{y} = \dfrac{4}{3} \cdot \dfrac{6}{1} \cdot \dfrac{2}{3} = \dfrac{16}{3}$.

8. **(E)** Since the center tile is on both diagonals, it follows that there are 51 black tiles on each diagonal. Thus, there are 51 tiles on each side for a total of

$$51^2 = (50+1)^2 = 50^2 + 2(50) + 1 = 2601$$

tiles on the floor.

9. **(E)** Partition the figure into 16 equilateral triangles as shown. Since each side of each of these 16 triangles has length

$$s = \frac{1}{2}\left(2\sqrt{3}\right) = \sqrt{3},$$

the total area is

$$16\left(s^2\frac{\sqrt{3}}{4}\right) = 16\left(3\frac{\sqrt{3}}{4}\right) = 12\sqrt{3}.$$

OR

In each of the two equilateral triangles at the ends, insert the segment connecting the midpoints of the outer side and the side on the line, as indicated. Observe that the area covered is enclosed by 5 congruent rhombi and 6 congruent equilateral triangles. The rhombi have diagonals with lengths 3 and $\sqrt{3}$, while the triangles have sides $\sqrt{3}$. Hence the area is

$$5\left(\frac{3\sqrt{3}}{2}\right) + 6\left(3\frac{\sqrt{3}}{4}\right) = 12\sqrt{3}.$$

OR

Since the five equilateral triangles overlap in four smaller triangles whose sides are half as large, the area is

$$5\left(12\frac{\sqrt{3}}{4}\right) - 4\left(3\frac{\sqrt{3}}{4}\right) = 12\sqrt{3}.$$

OR

The area covered consists of a trapezoid with height 3 and bases $6\sqrt{3}$ and $4\sqrt{3}$ minus four equilateral triangles whose sides and altitudes are half those of the given triangles. Therefore the answer is

$$3 \cdot \frac{6\sqrt{3} + 4\sqrt{3}}{2} - 4\left(3\frac{\sqrt{3}}{4}\right) = 12\sqrt{3}.$$

10. **(D)** Since $kx = 3k + 12$, it follows that $x = 3 + (12/k)$. Thus the equation has an integer solution if and only if k is a factor of 12.

Since $k > 0$, it follows that k is one of the six values 1, 2, 3, 4, 6 or 12.

OR

Since $k(x-3) = 12 = 12 \cdot 1 = 6 \cdot 2 = 4 \cdot 3 = 3 \cdot 4 = 2 \cdot 6 = 1 \cdot 12$, the 6 possible positive integer values of k are 12, 6, 4, 3, 2 and 1.

11. **(B)** Draw the radius from the center D of the circles to the point E where $\overline{BC}$ is tangent to the smaller circle. Since $\overline{DE} \perp \overline{BC}$, DEC and ABC are similar right triangles, so

$$\frac{DE}{AB} = \frac{CD}{CA} = \frac{1}{2}.$$

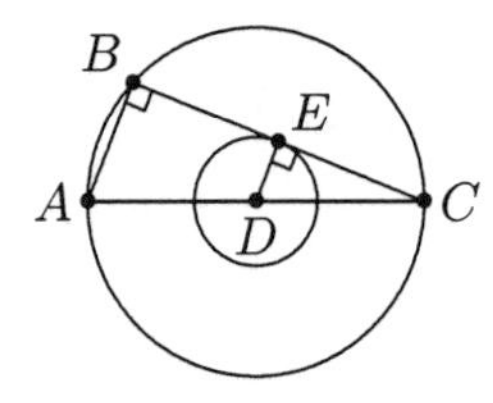

Thus the radius, DE, of the smaller circle is 6 since $AB = 12$. Hence the radius of the larger circle is 18 since the ratio of the radii is $1:3$.

OR

Label the diagram as above. Then

$$\sin C = \frac{AB}{AC} = \frac{DE}{DC} = \frac{1}{3}.$$

Hence, the radius of the larger circle is $AC/2 = 3AB/2 = 36/2 = 18$.

12. **(C)** The equation of the given line can be written as

$$y = \frac{1}{3}x + \frac{11}{3}.$$

The y-intercept, b, of the image is $-11/3$ and the slope, m, is $-1/3$. Thus $m + b = -4$.

OR

If the point (x, y) is on the reflection of the given line, then the point $(x, -y)$ is on the given line. Hence $x - 3(-y) + 11 = 0$, so $x + 3y = -11$ is an equation for the reflected line. The equation of this line can be written

$$y = -\frac{1}{3}x - \frac{11}{3},$$

so $m + b = -4$.

OR

On any line with slope m' and y-intercept b', $m'+b'$ is the y-coordinate when $x = 1$. Substitute $x = 1$ into the equation of given line:

$$1 - 3y + 11 = 0.$$

Solve to find that $y = 4$ when $x = 1$. The reflection of $(1, 4)$, namely $(1, -4)$, must be on the reflected line. Therefore, for the reflected line, $m + b = -4$.

Comment. This last solution explores a special case of the fact that the reflection of $y = mx + b$ over the x-axis is $y = -mx - b$.

13. **(C)** Multiply the numerator and denominator of the given fraction by ab to obtain

$$\frac{a^2b + a}{b + ab^2} = \frac{a(ab + 1)}{b(1 + ab)} = \frac{a}{b} = 13.$$

Thus $a = 13b$; and $a + b \leq 100$ implies $14b \leq 100$, so $0 < b \leq 7$. For each of the seven possible values of $b = 1, 2, 3, 4, 5, 6, 7$, the pair $(13b, b)$ is a solution.

14. **(E)** The graph of $y = x-2$ is a line. The graph of $y = (x^2-4)/(x+2)$ is almost a line, but there is a point at $x = -2$ missing. The graph of $(x + 2)y = x^2 - 4$ is a pair of intersecting lines, $x = -2$ and $y = x - 2$.

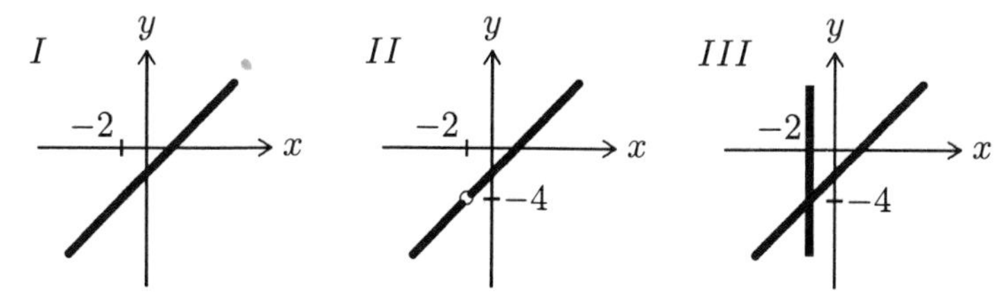

15. **(B)** We compute

$$z_1 = 0,\ z_2 = i,\ z_3 = i - 1,\ z_4 = -i,\ z_5 = i - 1.$$

Since $z_5 = z_3$, it follows that

$$z_{111} = z_{109} = z_{107} = \cdots = z_5 = z_3 = i - 1,$$

which is $\sqrt{2}$ units from the origin.

Note. The *Mandelbrot set* is defined to be the set of complex numbers c for which all the terms of the sequence defined by $z_1 = 0$, $z_{n+1} = z_n^2 + c$ for $n \geq 1$, stay close to the origin. Thus $c = i$ is in the Mandelbrot set.

16. **(E)** An easily-proved property of proportions is:

$$\text{If } \frac{a}{b} = \frac{c}{d} = \frac{e}{f} = k, \text{ then } \frac{a+c+e}{b+d+f} = k.$$

Use this to obtain

$$\frac{x}{y} = \frac{y+(x+y)+x}{(x-z)+z+y} = \frac{2x+2y}{x+y} = 2.$$

Note. In general, $x = 4t$, $y = 2t$ and $z = 3t$ for $t > 0$.

OR

Simplify

$$\frac{y}{x-z} = \frac{x}{y} \quad \text{and} \quad \frac{x+y}{z} = \frac{x}{y}$$

to obtain

$$y^2 = x^2 - xz \quad \text{and} \quad xy + y^2 = xz,$$

respectively. Substitute for xz to find that

$$y^2 = x^2 - (xy + y^2),$$

$$\text{so} \quad 0 = x^2 - xy - 2y^2 = (x-2y)(x+y).$$

Thus $x/y = 2$ or $x/y = -1$. Since $x, y > 0$, it follows that $x/y = 2$.

OR

Let $u = x/y$ and $v = z/y$. Then we are given that

$$\frac{1}{u-v} = \frac{u+1}{v} = u > 0.$$

Equate both fractional expressions to u to obtain

$$1 = u^2 - uv \qquad \text{and} \qquad u + 1 = uv.$$

$$\text{Hence} \qquad 1 = u^2 - (u+1),$$

$$\text{or} \qquad 0 = u^2 - u - 2 = (u-2)(u+1),$$

so $x/y = u = 2$ since $x, y > 0$.

Note. One can immediately eliminate three of the choices by noting that $x > z$ since $\dfrac{y}{x-z} = \dfrac{x}{y} > 0$. Hence $\dfrac{x+y}{z} > 1$, which means that $\dfrac{x}{y} > 1$.

17. **(B)** Since $0+1+2+\cdots+9=45$ and

$$N = 19\,\underbrace{20\,21\,\cdots\,29}_{10\cdot2+45=65}\,\underbrace{30\,31\,\cdots\,39}_{10\cdot3+45=75}\cdots\underbrace{80\,81\,\cdots\,89}_{10\cdot8+45=125}\,90\,91\,92,$$

the sum of the digits of N is

$$\begin{aligned} S &= (1+9)+65+75+\cdots+125+(3\cdot9+3) \\ &= (10)+\left(7\cdot\frac{65+125}{2}\right)+(30) \\ &= 705 = 9(78)+3. \end{aligned}$$

Thus S has a factor of 3 but not 9, so the highest power of 3 that is a divisor of N is 3^1 and $k=1$.

OR

The integer 3 [or 9] will divide N if and only if it divides the sum of 19, 20, ..., 92. (*Why?* See note below.) Since

$$19+20+\cdots+92 = 74\cdot\frac{19+92}{2} = 37\cdot111 = 37^2\cdot3,$$

it follows that $k=1$.

Note. Consider N as a 74-digit base-100 number with digits 19, 20, ..., 92. The sum of these digits is $3\cdot37^2$. The "casting out nines" procedure says:

- *"The greatest common factor of a positive integer and* 9 *is the same as the greatest common factor of the sum of the base* 10 *digits of the integer and* 9*."*

The proof of this procedure can be generalized to:

- *"The greatest common factor of a positive integer and* $b-1$ *is the same as the greatest common factor of the sum of the base* b *digits of the integer and* $b-1$*."*

Therefore, the greatest common factor of N and 99 equals the greatest common factor of $37^2\cdot3$ and 99.

18. **(D)** If $a_1=a$ and $a_2=b$ then $(a_3,\ a_4,\ a_5,\ a_6,\ a_7,\ a_8) =$

$$(a+b,\ a+2b,\ 2a+3b,\ 3a+5b,\ 5a+8b,\ 8a+13b).$$

Therefore $5a+8b = a_7 = 120$. Since $5a = 8(15-b)$ and 8 is relatively prime to 5, a must be a multiple of 8. Similarly, b must be a multiple of 5. Let $a = 8j$ and $b = 5k$ to obtain $40j + 40k = 120$, which has two solutions in positive integers, $(j,k) = (1,2)$ and $(2,1)$. When $(j,k) = 2,1)$, $(a,b) = (16,5)$, which is impossible since the sequence is increasing, so $(j,k) = (1,2)$ and $(a,b) = (8,10)$. Consequently, $a_8 = 8a + 13b = 194$.

Note. This sequence begins with the eight terms

$$8, 10, 18, 28, 46, 74, 120, 194.$$

19. **(D)** Let each edge of the cube be $2e$. Then the volume of the cube is $8e^3$. Each of the 8 tetrahedra removed has an isosceles right triangle of area $e^2/2$ as a base and an altitude of length e to this base. Hence the volume of the cuboctahedron is

$$8e^3 - 8\left(\frac{1}{3} \cdot e \cdot \frac{e^2}{2}\right) = \frac{20e^3}{3}.$$

The ratio of the volume of the cuboctahedron to the volume of the cube is

$$\frac{20e^3/3}{8e^3} = \frac{5}{6} = 83\frac{1}{3}\%.$$

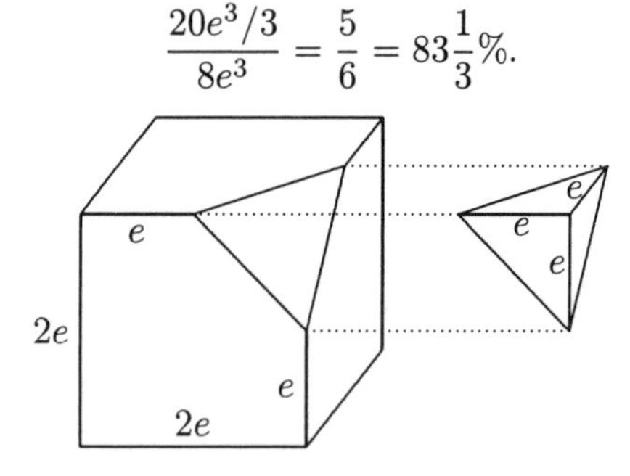

OR

Partition the cube with edge-length $2e$ by three planes parallel to the faces of the cube and through the center of the cube. In each of the resulting eight cubes, the same ratio r, of the volume it shares with the cuboctahedron to its volume, is

$$r = \frac{e^3 - (1/3)e(e^2/2)}{e^3} = \left(1 - \frac{1}{6}\right) e^3/e^3 = 83\frac{1}{3}\%,$$

and this is the same as the ratio requested in the problem.

20. **(D)** Partition the n-pointed regular star into the regular n-gon $B_1B_2\cdots B_n$ and n isosceles triangles congruent to $\triangle B_1A_2B_2$. The sum of the star's interior angles is the sum of the interior angles of the regular n-gon plus the sum of the interior angles of the n triangles, which is

$$(n-2)180^\circ + n180^\circ = (2n-2)180^\circ.$$

Since the interior angles of the star consist of n angles congruent to A_1 and n angles congruent to $360^\circ - B_1$,

$$(2n-2)180^\circ = n\angle A_1 + n(360^\circ - \angle B_1)$$
$$n(\angle B_1 - \angle A_1) = 2\cdot 180^\circ.$$

Since $\angle B_1 - \angle A_1 = 10^\circ$, $n = 36$.

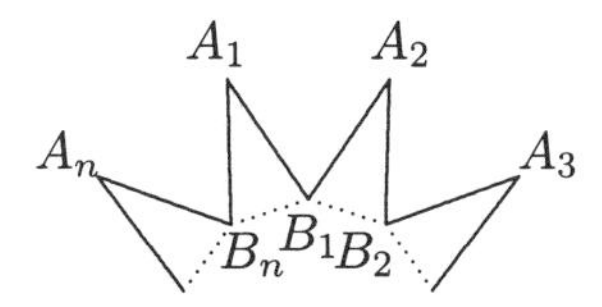

Note. In general, the sum of the interior angles of *any* N-sided simple closed polygon, convex or not, is $(N-2)180^\circ$.

OR

Extend $\overline{A_1B_1}$ and $\overline{A_2B_2}$ to their intersection C_1, $\overline{A_2B_2}$ and $\overline{A_3B_3}$ to their intersection $C_2, \ldots, \overline{A_nB_n}$ and $\overline{A_1B_1}$ to their intersection C_n. Note that the n triangles $A_1C_1A_2$, $A_2C_2A_3, \ldots,$ $A_nC_nA_1$ can be translated (moved without rotation) so all n points, C_i, coincide, and the angles with these vertices form a 360° angle. Since each $\angle C_i = 10^\circ$, $n \cdot 10^\circ = 360^\circ$, so $n = 36$.

OR

Let the measure of the acute angle at each A_i be A° and the measure of the acute angle at each B_i be B°. Let O be the center of the star. The $2n$ triangles A_iB_iO and $A_{i+1}B_iO$ (where $A_{n+1} = A_1$) are all congruent triangles whose angles measure

$$\frac{A^\circ}{2},\quad 180^\circ - \frac{B^\circ}{2} \quad\text{and}\quad \frac{360^\circ}{2n}.$$

Therefore

$$\frac{A^\circ}{2} + \left(180^\circ - \frac{B^\circ}{2}\right) + \frac{360^\circ}{2n} = 180^\circ$$

or

$$\frac{360^\circ}{2n} = \frac{B^\circ - A^\circ}{2} = \frac{10^\circ}{2},$$

so $n = 36$.

OR

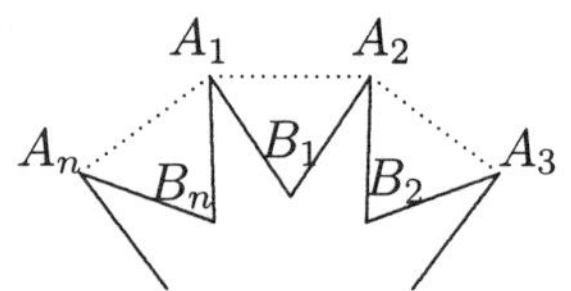

Draw the segments $\overline{A_1A_2}$, $\overline{A_2A_3}$, …, $\overline{A_nA_1}$ to form a regular n-gon. In $\triangle A_1A_2B_1$, $\angle B_1A_1A_2 = \angle B_1A_2A_1$, so

$$2\angle B_1A_1A_2 + \angle A_1B_1A_2 = 180^\circ. \tag{1}$$

The measure of an interior angle of regular polygon $A_1A_2\cdots A_n$ is $(n-2)180^\circ/n$, so

$$\begin{aligned}\angle A_1A_2B_1 + \angle B_1A_2B_2 + \angle B_2A_2A_3 &= \angle A_1A_2A_3,\\ 2\angle B_1A_1A_2 + \angle B_1A_2B_2 &= \frac{n-2}{n}180^\circ. \qquad (2)\end{aligned}$$

Subtract equation (2) from equation (1):

$$\angle A_1B_1A_2 - \angle B_1A_2B_2 = 180^\circ - \left(\frac{n-2}{n}180^\circ\right) = \frac{2}{n}\cdot 180^\circ.$$

But $\angle A_1B_1A_2 - \angle B_1A_2B_2 = 10^\circ$, so $n = 36$.

21. **(A)** Since $\dfrac{S_1 + S_2 + \cdots + S_{99}}{99} = 1000$, it follows that

$$S_1 + S_2 + \cdots + S_{99} = 99000.$$

Thus

$$\begin{aligned}\frac{1 + (1+S_1) + (1+S_2) + \cdots + (1+S_{99})}{100} &= \frac{100 + (S_1 + S_2 + \cdots + S_{99})}{100}\\ &= \frac{100 + 99000}{100} = 991.\end{aligned}$$

OR

In general, let C be the Cesàro sum of the n-term sequence $(b_1, b_2, \ldots, b_n)$. Then

$$b_1 + (b_1 + b_2) + (b_1 + b_2 + b_3) + \cdots + (b_1 + b_2 + \cdots + b_n) = nC.$$

The Cesàro sum of the $(n+1)$-term sequence $k, b_1, b_2, \ldots, b_n$ is then

$$\frac{k + (k + b_1) + (k + b_1 + b_2) + \cdots + (k + b_1 + b_2 + \cdots + b_n)}{n+1}$$
$$= \frac{(n+1)k + (b_1 + (b_1 + b_2) + \cdots + (b_1 + b_2 + \cdots + b_n))}{n+1}$$
$$= \frac{(n+1)k + nC}{n+1} = k + \frac{n}{n+1}C.$$

In the problem, we have $n = 99$, $C = 1000$, and $k = 1$, which gives a Cesàro sum of 991.

Note. Cesàro sums are named after the nineteenth century mathematician E. Cesàro. Cesàro sums arise in many areas of mathematics such as in the study of Fourier Series.

22. **(B)** A point of intersection in the first quadrant is obtained whenever two of the segments cross to form an $\times$. An $\times$ is uniquely determined by selecting two of the points on $\mathbf{X}^+$ and two of the points on $\mathbf{Y}^+$. The maximum number of these intersections is obtained by selecting the points on $\mathbf{X}^+$ and $\mathbf{Y}^+$ so that no three of the 50 segments intersect in the same point. Therefore, the maximum number of intersections is $\binom{10}{2}\binom{5}{2} = 45 \cdot 10 = 450$.

OR

Choose ten points, $x_1 < x_2 < \ldots < x_{10}$, on $\mathbf{X}^+$. Choose y_1 on $\mathbf{Y}^+$ and draw the ten segments joining y_1 to the ten points on $\mathbf{X}^+$. Choose $y_2 > y_1$ on $\mathbf{Y}^+$, and note that, as the segments $\overline{y_2x_1}$, $\overline{y_2x_2}$, …, $\overline{y_2x_{10}}$ are drawn, $9 + 8 + 7 + \cdots + 0 = 45$ intersections are formed. Choose $y_3 > y_2$ on $\mathbf{Y}^+$ so no segment $\overline{y_3x_i}$ goes through a previously counted intersection, and note that $2(9+8+7+\cdots+0) = 2{\cdot}45$ new intersections are formed. Similarly, for judiciously chosen y_4 and y_5 on $\mathbf{Y}^+$, one can generate at most $3 \cdot 45$ and $4 \cdot 45$ new intersections, respectively. Hence, the maximum number of intersections is $(1{+}2{+}3{+}4)45 = 450$ intersections.

23. **(E)** Partition $F = \{1, 2, 3, \ldots, 50\}$ into seven subsets, F_0, F_1, ..., F_6, so that all the elements of F_i leave a remainder of i when divided by 7:

$$F_0 = \{7, 14, 21, 28, 35, 42, 49\},$$
$$F_1 = \{1,\ \ 8, 15, 22, 29, 36, 43, 50\},$$
$$F_2 = \{2,\ \ 9, 16, 23, 30, 37, 44\},$$
$$F_3 = \{3, 10, 17, 24, 31, 38, 45\},$$
$$F_4 = \{4, 11, 18, 25, 32, 39, 46\},$$
$$F_5 = \{5, 12, 19, 26, 33, 40, 47\},$$
$$F_6 = \{6, 13, 20, 27, 34, 41, 48\}.$$

Note that S can contain at most one member of F_0, but that if S contains some member of any of the other subsets, then it can contain all of the members of that subset. Also, S cannot contain members of both F_1 and F_6, or both F_2 and F_5, or both F_3 and F_4. Since F_1 contains 8 members and each of the other subsets contains 7 members, the largest subset, S, can be constructed by selecting one member of F_0, all the members of F_1, all the members of either F_2 or F_5, and all of the members of either F_3 or F_4. Thus the largest subset, S, contains $1+8+7+7 = 23$ elements.

24. **(C)** Since the area $[ABCD] = 10$ and $AD = BC = 5$, the distance between lines $\overline{AD}$ and $\overline{BC}$ is $10/5 = 2$. Thus, if the altitude of $\triangle AEG$ from E is h, then the altitude of $\triangle BEF$ from E is $2-h$. Since $AG = BF = 2$,

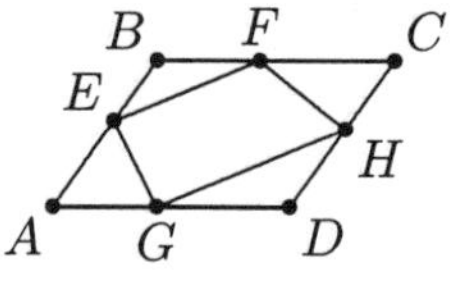

$$[AEG] + [BEF] = \frac{1}{2}AG \cdot h + \frac{1}{2}BF \cdot (2-h) = h + (2-h) = 2.$$

Similarly, since the sum of the altitudes of triangles CFH and DGH from H is 2 and $CF = DG = 3$, it follows that $[CFH]+[DGH] = 3$. Hence

$$[EFHG] = [ABCD] - ([AEG] + [BEF] + [CFH] + [DGH])$$
$$= 10 - (2+3) = 5.$$

OR

Note that $ABFG$ is a parallelogram. Hence $[EFG] = \frac{1}{2}[ABFG]$, and similarly, $[HFG] = \frac{1}{2}[CDGF]$. Consequently,

$$[EFHG] = \frac{1}{2}[ABCD] = 5.$$

Note. Not only is the choice of E and H completely arbitrary on their respective segments, but F and G can be chosen as any two points on $\overline{BC}$ and $\overline{AD}$ as long as $BF = AG$.

OR

Let $\angle DAB = \alpha$. Then

$$[ABCD] = 10 = 3 \cdot 5 \sin \alpha, \quad \text{so} \quad \sin \alpha = \frac{2}{3}.$$

Since the length of the altitude from G in $\triangle AEG$ is $AG \sin \alpha$ and that from F in $\triangle BEF$ is $BF \sin \alpha$,

$$[AEG] = \frac{1}{2} \cdot 2 \cdot \left(2 \cdot \frac{2}{3}\right) = \frac{4}{3} \quad \text{and} \quad [BEF] = \frac{1}{2} \cdot 1 \cdot \left(2 \cdot \frac{2}{3}\right) = \frac{2}{3}.$$

Since $\overline{GH} \parallel \overline{EF}$, $\triangle BEF \sim \triangle DHG$, and therefore

$$\frac{DH}{BE} = \frac{DG}{BF} \quad \text{so} \quad DH = \frac{BE \cdot DG}{BF} = \frac{1 \cdot 3}{2}$$

$$\text{and} \quad HC = DC - DH = 3 - \frac{3}{2} = \frac{3}{2}.$$

Therefore,

$$[DGH] = [CFH] = \frac{1}{2} FC \cdot HC \sin \alpha = \frac{1}{2} \cdot 3 \cdot \frac{3}{2} \cdot \frac{2}{3} = \frac{3}{2},$$

and

$$\begin{aligned}[EFHG] &= [ABCD] - [AEG] - [BEF] - [DGH] - [CFH] \\ &= 10 - \frac{4}{3} - \frac{2}{3} - 2\left(\frac{3}{2}\right) = 5.\end{aligned}$$

25. **(E)** Extend $\overline{CB}$ and $\overline{DA}$ to meet at E. Since $\angle E = 30°$, $EB = 6$. Hence $EC = 10$ and $CD = 10/\sqrt{3}$. In general, whenever $\angle ABC = 120°$, if $AB = x$ and $BC = y$, then $EB = 2x$ and $EC = 2x + y$, so $CD = (2x + y)/\sqrt{3}$.

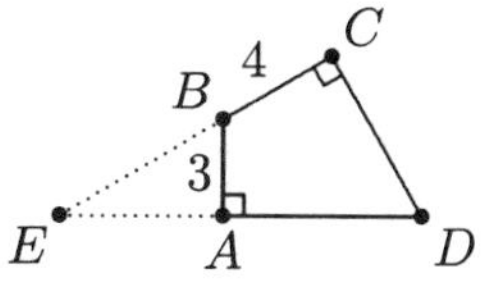

C
B
4
3
E
A
D

OR

Draw a line through B parallel to $\overline{AD}$ intersecting CD at H. Then drop a perpendicular from H to I on $\overline{AD}$. Note that BHC and HDI are both 30°-60°-90° triangles. Thus $CH = 4/\sqrt{3}$. Since $HI = AB = 3$, it follows that $HD = 6/\sqrt{3}$. Hence

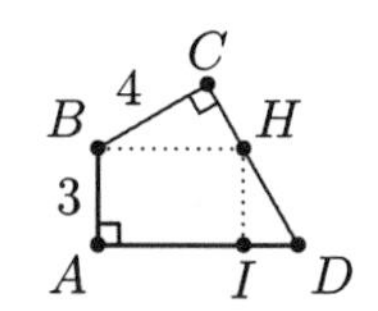

$$CD = CH + HD = \frac{4}{\sqrt{3}} + \frac{6}{\sqrt{3}} = \frac{10}{\sqrt{3}}.$$

OR

Draw a line through C parallel to $\overline{AD}$ and meeting $\overline{AB}$ extended at F. Then $\angle BCF = 30°$, so $BF = 2$. Drop a perpendicular from C to G on $\overline{AD}$. Since $AFCG$ is a rectangle, $CG = AF = AB + BF = 5$. Since $\angle CDG = 60°$, we have $CD = (2/\sqrt{3})CG = 10/\sqrt{3}$.

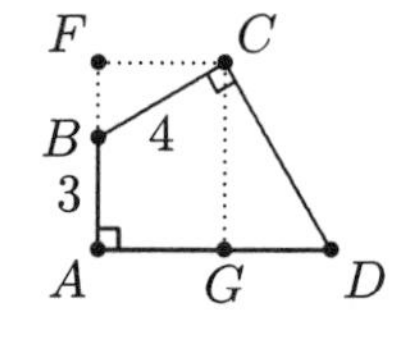

OR

Let $AD = x$ and $CD = y$. By the law of cosines applied to triangles ABC and ADC,

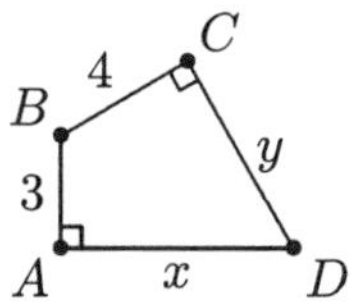

$$x^2 + y^2 - 2{\cdot}xy\frac{1}{2} = AC^2$$

$$= 3^2 + 4^2 + 2 \cdot 3 \cdot 4 \cdot \frac{1}{2} = 37.$$

This simplifies to

$$xy = x^2 + y^2 - 37. \qquad (1)$$

Apply the Pythagorean theorem to triangles BAD and BCD to find

$$x^2 + 9 = BD^2 = y^2 + 16,$$

$$\text{or} \quad 0 = x^2 - y^2 - 7. \qquad (2)$$

Subtract (2) from (1) and divide by y to find

$$x = 2y - \frac{30}{y}$$

which, upon substitution into (2), leads to

$$0 = 3y^2 - 127 + \frac{900}{y^2} = \left(3 - \frac{100}{y^2}\right)\left(y^2 - 9\right).$$

Solve to find $y = \pm 3, \pm 10/\sqrt{3}$. Since $CD > 3$, the only applicable root is $CD = y = 10/\sqrt{3}$.

OR

Since opposite angles in $ABCD$ are supplementary, this quadrilateral can be inscribed in a circle with center O. Apply the law of cosines to $\triangle ABC$ to obtain $AC = \sqrt{37}$ as in the previous solution. Then apply the law of cosines to $\triangle AOC$ to find the radius of the circle $r = OA = OC$:

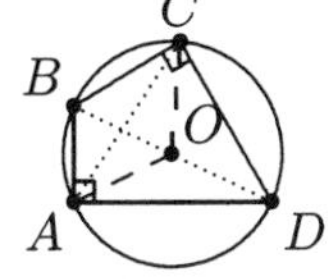

$$AO^2 + OC^2 - 2(AO)(OC)\cos 120^\circ = AC^2;$$

$$r^2 + r^2 + r^2 = 37.$$

Since $r = \sqrt{37/3}$, it follows that the diameter of the circle, $BD = 2\sqrt{37/3}$. By the Pythagorean theorem,

$$CD = \sqrt{BD^2 - BC^2}$$
$$= \sqrt{4\cdot\frac{37}{3} - 4^2} = \frac{10}{\sqrt{3}}.$$

OR

Let $\angle CBD = \theta$. Then $\angle ABD = 120^\circ - \theta$. In $\triangle CBD$ we have $\cos\theta = 4/BD$ and $\sin\theta = CD/BD$. In $\triangle ABD$ we have

$$\frac{3}{BD} = \cos(120^\circ - \theta)$$
$$= (\cos 120^\circ)(\cos\theta) + (\sin 120^\circ)(\sin\theta)$$
$$= -\frac{1}{2}\cdot\frac{4}{BD} + \frac{\sqrt{3}}{2}\cdot\frac{CD}{BD}.$$

Multiply by $2BD$ to obtain $6 = -4 + CD\sqrt{3}$ which we solve to obtain $CD = 10/\sqrt{3}$.

26. **(B)** Since $\overline{CD} \perp \overline{AB}$ and $AC = DC = BC = 1$, it follows that $\triangle ACD$ and $\triangle BCD$ are isosceles right triangles, so $\angle BAD = \angle ABD = 45^\circ$ and $BD = \sqrt{2}$. Note that $\angle EDF = \angle ADB = 90^\circ$ and $DE = BE - BD = 2 - \sqrt{2}$. Since the sectors ABE and BAF

are congruent, the area of the "smile" is

$$\begin{aligned}
&\text{area(sector } EDF) + 2\,[\text{area(sector } ABE)] \\
&\qquad\qquad - \text{area}(\triangle ABD) - \text{area(semicircle } ADB) \\
&= \frac{90}{360}\pi\left(2-\sqrt{2}\right)^2 + 2\left[\frac{45}{360}\pi(2)^2\right] - \frac{1}{2}\cdot 2\cdot 1 - \frac{1}{2}\pi(1)^2 \\
&= \left(\frac{3}{2}-\sqrt{2}\right)\pi + 2\left(\frac{\pi}{2}\right) - 1 - \frac{\pi}{2} \\
&= 2\pi - \pi\sqrt{2} - 1.
\end{aligned}$$

27. **(D)** Using properties of secant segments (*power of a point*), it follows that

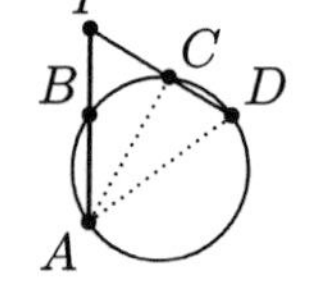

$$PD \cdot PC = PA \cdot PB = 18 \cdot 8 = 144.$$

Use

$$PD \cdot PC = 144 \quad \text{and} \quad PD = PC + 7$$

to find that $PD = 16$ and $PC = 9$. Since $AP = 2PC$ and $\angle APC = 60^\circ$, it follows that $\angle ACP = 90^\circ$. Thus $AC = PC\sqrt{3} = 9\sqrt{3}$. Since $\angle ACD$ is also a right angle, $\overline{AD}$ is a diameter of the circle. Apply the Pythagorean Theorem to triangle ACD to obtain

$$(2r)^2 = AD^2 = AC^2 + CD^2 = 3 \cdot 9^2 + 7^2 = 292,$$

from which it follows that $r^2 = 73$.

28. **(B)** The quadratic formula† leads to the roots

$$z = \frac{1}{2}\left(1 \pm \sqrt{21 - 20i}\right).$$

To find $\sqrt{21-20i}$, let $(a+bi)^2 = 21 - 20i$ where a and b are real. Equating real and imaginary parts leads to $a^2 - b^2 = 21$ and $2ab = -20$. Solve these equations simultaneously:

$$a^2(a^2-21) = a^2b^2 = \left(\frac{-20}{2}\right)^2 = 100;$$

$$a^4 - 21a^2 - 100 = 0;$$

† Many times we might think only of real numbers as we complete the square and derive the quadratic formula. All the operations involved in this derivation are also valid within the complex number field.

$$(a^2-25)(a^2+4)=0;$$

$$a^2-25=0; \qquad a=\pm 5, \text{ so } b=\mp 2.$$

Thus $a+bi=5-2i$ or $-5+2i$. Therefore

$$z=\frac{1}{2}\left[1\pm(5-2i)\right]=\begin{cases}3-i & \text{or}\\ -2+i.\end{cases}$$

The product of the real parts of these two roots is $3(-2)=-6$.

Note. One could also use the equation $a^2-b^2=21$ together with $a^2+b^2=|a+bi|^2=|21-20i|=29$ and solve simultaneously to obtain $2a^2=50$, from which it follows that $a=\pm 5$.

OR

Let $z=x+iy$ where x and y are real. Therefore

$$(x+iy)^2-(x+iy)=5-5i.$$

Equate real and imaginary parts to find that

$$x^2-y^2-x=5 \qquad \text{and} \qquad 2xy-y=-5.$$

To solve the last equation for y in terms of x first verify that $x\neq 1/2$:

If $x=1/2$, then

$$5=x^2-y^2-x\le\left(\frac{1}{2}\right)^2-0-\frac{1}{2}<0,$$

a contradiction.

Hence

$$y=\frac{5}{1-2x} \qquad \text{so} \qquad x^2-\left(\frac{5}{1-2x}\right)^2-x=5. \qquad (*)$$

Let $f(x)=4x^4-8x^3-15x^2+19x-30$ and note that $(*)$ simplifies to $f(x)=0$. By Descartes' law of signs, $f(x)=0$ has one negative solution. Some negative integers which may be solutions are -1, -2, -3, ... Testing these we find that $f(-2)=0$, so $x=-2$ is the one negative solution. Let

$$g(x)=\frac{f(x)}{x+2}=4x^3-16x^2+17x-15.$$

Possible positive integer solutions to $g(x)=0$ are 1, 3, 5, 15. Testing these we find that $f(3)=g(3)=0$, so $x=3$ is another possible solution. Use the equation for y in $(*)$, and verify that $-2+i$ and

$3-i$ are the two roots of the original equation. The product of their real parts is -6.

29. **(D)** Let p be the probability that the total number of heads is even, and let q be the probability that the total number of heads is odd. Since the probability of tossing k heads and $(50-k)$ tails is

$$\binom{50}{k}\left(\frac{2}{3}\right)^{k}\left(\frac{1}{3}\right)^{50-k},$$

we have

$$p=\binom{50}{0}\left(\frac{2}{3}\right)^{0}\left(\frac{1}{3}\right)^{50}+\binom{50}{2}\left(\frac{2}{3}\right)^{2}\left(\frac{1}{3}\right)^{48}+\cdots+\binom{50}{50}\left(\frac{2}{3}\right)^{50}\left(\frac{1}{3}\right)^{0}$$

and

$$q=\binom{50}{1}\left(\frac{2}{3}\right)^{1}\left(\frac{1}{3}\right)^{49}+\binom{50}{3}\left(\frac{2}{3}\right)^{3}\left(\frac{1}{3}\right)^{47}+\cdots+\binom{50}{49}\left(\frac{2}{3}\right)^{49}\left(\frac{1}{3}\right)^{1}.$$

Simplify $p-q$ using the Binomial Theorem:

$$p-q=\sum_{k=0}^{50}(-1)^k\binom{50}{k}\left(\frac{2}{3}\right)^{k}\left(\frac{1}{3}\right)^{50-k}=\left(\frac{2}{3}-\frac{1}{3}\right)^{50}=\frac{1}{3^{50}}.$$

Since $p+q=1$, solve

$$p+q=1$$
$$p-q=\frac{1}{3^{50}}$$

simultaneously to get

$$p=\frac{1}{2}\left(1+\frac{1}{3^{50}}\right).$$

OR

Instead of flipping a biased two-sided coin, we could toss an unbiased three-sided die (such as an equilateral triangle in the plane) where the sides were labeled "HEAD", "HEAD", and "TAIL". Equivalently, one could consider all n-digit numbers in base 3 (with leading 0's

permitted) and pick one of these at random. Think of a 0 as a tail and a non-zero digit as a head. Then our question becomes: What is the probability that the randomly chosen 50-digit number has an even number of non-zero digits?
Consider a few easy cases. In these examples, the base 3 numbers with an even number of non-zero digits are underlined:

$$\underline{00}, 01, 02, 10, \underline{11}, \underline{12}, 20, \underline{21}, \underline{22}$$

$$000, \underline{001}, \underline{002}, \underline{010}, 011, 012, \underline{020}, 021, 022,$$

$$\underline{100}, 101, 102, 110, \underline{111}, \underline{112}, 120, \underline{121}, \underline{122},$$

$$\underline{200}, 201, 202, 210, \underline{211}, \underline{212}, 220, \underline{221}, \underline{222},$$

Thus, $5/9$ of the two-digit numbers contain an even number of non-zero digits, and $14/27$ of the three-digit numbers contain an even number of non-zero digits. We are led to conjecture that the probability is very close to, and slightly larger than, $1/2$. In fact, the following is true for even n:

Theorem. *Let n be even. Among all non-zero n-digit numbers written in base 3, exactly half of them have an even number of non-zero digits.*

Proof. We will give an explicit one-to-one correspondence between those base 3 numbers with an even number of non-zero digits and those with an odd number of non-zero digits.

Let x be a non-zero n-digit number written in base 3. Since n is even, we may group the digits in pairs. Since x is not zero, there must be some pair that is not '00'. Find the left-most pair that is not '00' and match this with the number y, formed by replacing that pair according to the following exchange rules:

$$01 \leftrightarrow 21$$

$$02 \leftrightarrow 12$$

$$10 \leftrightarrow 11$$

$$20 \leftrightarrow 22.$$

For example,

$$x = 00\,00\,\underline{02}\,01\,21\,12 \to y = 00\,00\,\underline{12}\,01\,21\,12$$

and $$y = 00\,00\,\underline{12}\,01\,21\,12 \to x = 00\,00\,\underline{02}\,01\,21\,12$$

Since y will map back into x by this rule if x maps into y, this is a one to one correspondence. Since each exchange rule either deletes or inserts one zero, the numbers with an odd number of non-zero digits are in one to one correspondence with the numbers with an even number of non-zero digits.

Thus, when n is even, there are 3^n n-digit numbers in base 3 and, of the $3^n - 1$ non-zero numbers, half have an even number of non-zero digits and half have an odd number. Since n is even, the number zero contains an even number of zeros. Hence, the probability of choosing a number with an even number of non-zero digits is

$$\frac{1+(3^n-1)/2}{3^n} = \frac{3^n+1}{2\cdot 3^n} = \frac{1}{2}\left(1+\frac{1}{3^n}\right)$$

for even n. Use $n = 50$ to find our answer.

Challenges. (1) Show that the theorem is false for odd n.
(2) Show that first grouping the digits into pairs (and hence that n is even) is necessary for the proof of the theorem by showing that applying the exchange rules to the leftmost adjacent digits that are not 00 does not lead to a one-to-one correspondence when $n = 3$.
(3) Find a formula for the number of n-digit non-zero numbers with an even number of non-zero digits when n is odd, and construct a proof of your formula by establishing some one to one correspondence.
(4) Let p_n be the probability that a randomly chosen n-digit number contains an even number of non-zero digits. The fact that

$$p_2 = \frac{5}{9} \quad\text{and}\quad p_3 = \frac{14}{27} = \frac{5+9}{3\cdot 9}$$

suggests that if $p_n = a/b$, then it might be the case that

$$p_{n+1} = \frac{a+b}{3b}.$$

Is this valid for all $n > 1$?

30. **(B)** Since $ABCD$ is isosceles, the center of the circle, P, must be the midpoint of $\overline{AB}$. When $x = m$, the circle must be tangent to $\overline{AD}$ at D and to $\overline{BC}$ at C. (*Why?*) Let Q be the foot of the perpendicular from D to $\overline{AB}$. Then $\triangle ADP$ is a right triangle with hypotenuse $\overline{AP}$, and $\overline{DQ}$ is its altitude to the hypotenuse. Since $\triangle ADQ \sim \triangle APD$,

$$\frac{AD}{AP} = \frac{AQ}{AD}, \quad\text{so}\quad m^2 = AD^2 = AQ\cdot AP = \frac{73}{2}\cdot\frac{92}{2} = 1679.$$

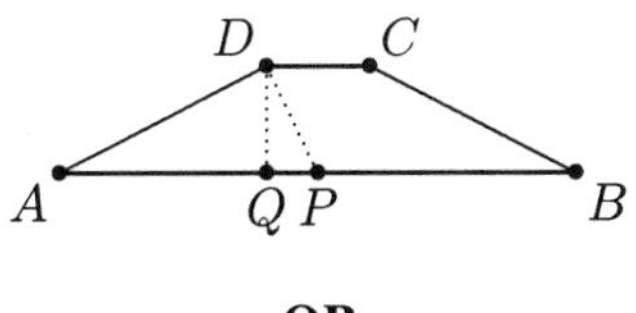

OR

Use analytic geometry, taking

$$A = (0,0), \quad P = (46,0), \quad D = \left(\frac{73}{2}, h\right).$$

We must minimize

$$AD^2 = \left(\frac{73}{2}\right)^2 + h^2$$

subject to "the line through P perpendicular to $\overline{AD}$ intersects segment $\overline{AD}$." The equations of the lines $\overline{AD}$ and the line through P perpendicular to $\overline{AD}$ are, respectively,

$$y = \frac{h}{73/2}x$$

$$\text{and} \qquad y = -\frac{73/2}{h}(x - 46).$$

The x-coordinate of the intersection of these lines is

$$x = \frac{73^2 \cdot 46}{4h^2 + 73^2}.$$

The minimal value of $AD^2 = (73/2)^2 + h^2$ occurs when h is as small as possible. Since $x = (73^2 \cdot 46)/(4h^2 + 73^2)$, it follows that when h is as small as possible x is as large as possible. But, for the intersection to occur on segment $\overline{AD}$, $x \le 73/2$. Therefore, let $x = 73/2$ and solve

$$\frac{73}{2} = \frac{73^2 \cdot 46}{4h^2 + 73^2}$$

to find that $h^2 = 73 \cdot 19/4$, and hence

$$AD^2 = \left(\frac{73}{2}\right)^2 + h^2 = \frac{73^2}{4} + \frac{73 \cdot 19}{4} = 1679.$$

44 AHSME Solutions

1. **(D)** $\boxed{1,-1,2} = 1^{-1} - (-1)^2 + 2^1 = 1 - 1 + 2 = 2.$

2. **(D)** We have $\angle B = 180^\circ - (55^\circ + 75^\circ) = 50^\circ$. Since $\triangle BDE$ is isosceles, $\angle BED = (180^\circ - \angle B)/2 = (180 - 50)^\circ/2 = 65^\circ$.

OR

Since $\angle B$ is in both triangles ABC and DBC, it follows that

$$\angle BDE + \angle BED = \angle BAC + \angle BCA = 130^\circ.$$

Since $BD = BE$, it follows that $\angle BED = \angle BDE$, so $\angle BED = 130^\circ/2 = 65^\circ$.

3. **(E)** Simplify by first writing as products of prime factors:

$$\frac{15^{30}}{45^{15}} = \frac{3^{30}5^{30}}{3^{15}3^{15}5^{15}} = 3^{30-15-15}5^{30-15} = 3^0 5^{15} = 5^{15}.$$

OR

Simplify by first grouping like powers, numerator and denominator:

$$\frac{15^{30}}{45^{15}} = 15^{15}\left(\frac{15}{45}\right)^{15} = 15^{15}\left(\frac{1}{3}\right)^{15} = \left(\frac{15}{3}\right)^{15} = 5^{15}.$$

OR

Eliminate all choices except **(C)** and **(E)** by noting that the prime 3 appears 30 times in the numerator and $2 \cdot 15$ times in the denominator.

Then eliminate choice **(C)** since the prime 5 appears more in the numerator than in the denominator.

4. **(E)** Substituting 3 for x yields $3 \circ y = 12 - 3y + 3y = 12$. Thus $3 \circ y = 12$ is true for all real numbers y.

5. **(A)** Last year the bicycle and helmet cost $\$160 + \$40 = \$200$. This year the cost of the bicycle increased $0.05(\$160) = \$8$, while the cost of the helmet increased $0.10(\$40) = \$4$. Thus it costs \$12 more for the bicycle and helmet this year. This is an increase of $12/200 = 6/100 = 6\%$.

6. **(B)** Begin by writing each term as a power of the prime 2:

$$\sqrt{\frac{8^{10}+4^{10}}{8^4+4^{11}}} = \sqrt{\frac{(2^3)^{10}+(2^2)^{10}}{(2^3)^4+(2^2)^{11}}} = \sqrt{\frac{2^{30}+2^{20}}{2^{12}+2^{22}}}$$

$$= \sqrt{\frac{2^{20}(2^{10}+1)}{2^{12}(1+2^{10})}} = \sqrt{2^8} = 2^4 = 16.$$

OR

Eliminate all but the correct choice using estimation:

$$\sqrt{\frac{8^{10}+4^{10}}{8^4+4^{11}}} = \sqrt{\frac{2^{30}+2^{20}}{2^{12}+2^{22}}} \approx \sqrt{\frac{2^{30}}{2^{22}}} = \left(2^8\right)^{1/2} = 16.$$

No other choice is even close!

7. **(E)** Note that $9R_k$ is represented by a sequence of k nines. Since

$$\frac{R_{24}}{R_4} = \frac{9R_{24}}{9R_4} = \frac{10^{24}-1}{10^4-1} = \frac{\left(10^4\right)^6-1}{10^4-1}$$

$$= 10^{20}+10^{16}+10^{12}+10^8+10^4+1$$

$$= 100010001000100010001,$$

there are $5 \times 3 = 15$ zeros in the quotient.

OR

Divide to compute the quotient:

$$\begin{array}{r} 1\;0001\;0001\;0001\;0001\;0001 \\ 1111 \overline{)1111\;1111\;1111\;1111\;1111\;1111} \end{array}$$

OR

Factor R_4 out of R_{24}:

$$\begin{aligned}\frac{R_{24}}{R_4} &= \frac{1111 \times 100010001000100010001}{1111} \\ &= 100010001000100010001.\end{aligned}$$

OR

This solution can be seen to be an application of the distributive law:

$$\begin{aligned}\frac{R_{24}}{R_4} =&[(1111\times 10^{20}) + (1111\times 10^{16}) + (1111\times 10^{12}) \\ &+ (1111\times 10^{8}) + (1111\times 10^{4}) + 1111]/1111 \\ =&\frac{1111[10^{20} + 10^{16} + 10^{12} + 10^{8} + 10^{4} + 1]}{1111} \\ =&100010001000100010001.\end{aligned}$$

8. **(D)** There are six such circles. Two enclose both C_1 and C_2, two enclose neither, and two enclose exactly one of C_1 and C_2.

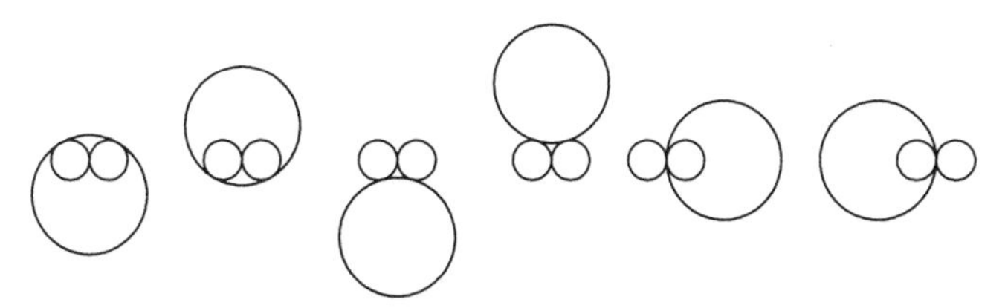

Query. What if the radius of the third circle were 2? $1\frac{1}{2}$? $\frac{1}{2}$?

9. **(D)** Let P be the world's population and W be its wealth. Then the $Pc/100$ citizens of $\mathcal{A}$ together own $Wd/100$ units of wealth, so each citizen of $\mathcal{A}$ owns

$$\frac{Wd/100}{Pc/100} = Wd/Pc$$

units of wealth. Similarly, each citizen of $\mathcal{B}$ owns Wf/Pe units of wealth. The required ratio is therefore

$$\frac{Wd/Pc}{Wf/Pe} = \frac{de}{cf}.$$

10. **(C)** Since

$$r = (3a)^{3b} = \left((3a)^3\right)^b = (27a^3)^b, \quad \text{and} \quad r = a^b x^b = (ax)^b,$$

we have $(27a^3)^b = (ax)^b$. Thus $27a^3 = ax$, which we solve to obtain $x = 27a^2$. To show that none of the other choices is correct, let $a = b = 1$.

11. **(A)** Use the definition of a logarithm to the base 2 repeatedly:

$$\begin{aligned}\log_2(\log_2(\log_2(x))) = 2 &\iff \log_2(\log_2(x)) = 2^2 = 4,\\ &\iff \log_2(x) = 2^4 = 16\\ &\iff x = 2^{16}.\end{aligned}$$

Since $2^{16} = 2^6 \cdot 2^{10} = 64 \cdot 1024$, it follows that x has 5 base-ten digits.

Comment. The estimation of the size of 2^{16} can also be obtained from

$$2^{16} = 2\left(2^5\right)^3 = 2 \cdot 32^3.$$

12. **(E)** Since we are given the value of f at twice a number, we must express $f(x)$ as $f\left(2\left(\frac{x}{2}\right)\right)$:

$$2f(x) = 2f\left(2 \cdot \frac{x}{2}\right) = 2\left(\frac{2}{2 + \frac{x}{2}}\right) = 2\left(\frac{4}{4 + x}\right) = \frac{8}{4 + x}.$$

13. **(D)** The inscribed square will touch each edge of the larger square, dividing that edge into segments x and y units long, where

$$x \leq y, \; x + y = 7 \quad \text{and} \quad x^2 + y^2 = 25.$$

Solve simultaneously to find $x = 3$ and $y = 4$. The greatest distance is therefore

$$\begin{aligned}AB &= \sqrt{y^2 + (x+y)^2}\\ &= \sqrt{4^2 + 7^2} = \sqrt{65}.\end{aligned}$$

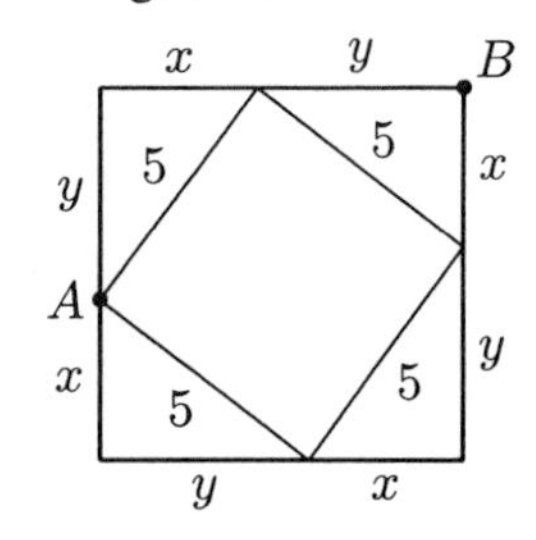

OR

The inscribed square will touch each edge of the larger square, dividing that edge into segments x and $7 - x$ units long, where $x \leq 7/2$. Apply

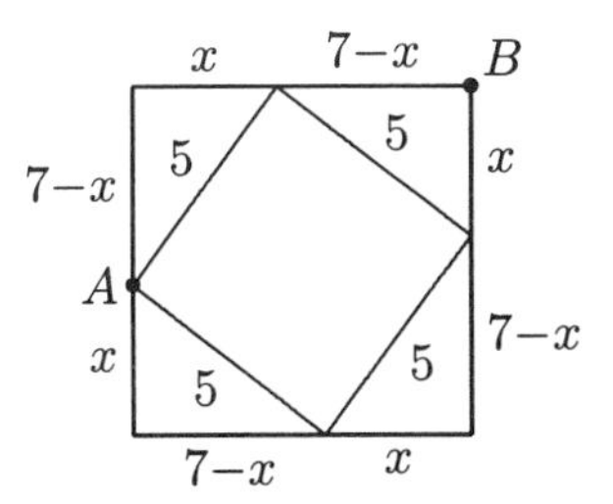

the *Pythagorean Theorem* to one of the right triangles inside the large square and outside the small one:

$$\begin{aligned} x^2 + (7-x)^2 &= 25 \\ x^2 + 49 - 14x + x^2 &= 25 \\ 2x^2 - 14x + 24 &= 0 \\ 2(x^2 - 7x + 12) &= 0 \quad \Rightarrow \quad x = 3,\ 7 - x = 4. \end{aligned}$$

$$\begin{aligned} AB &= \sqrt{(7-x)^2 + (x + (7-x))^2} \\ &= \sqrt{4^2 + 7^2} = \sqrt{65}. \end{aligned}$$

Note. The 3-4-5 triangle used in this problem can be generalized to any $a, b, c > 0$ with $a^2 + b^2 = c^2$. The problem would ask the same question about a square c units on a side inscribed in a square $a + b$ units on a side. The answer would be $b^2 + (a+b)^2$ where $a \le b$.

14. **(B)** Draw $\overline{CE}$. Since $EA = BC$ and $\angle A = \angle B$, it follows that $ABCE$ is an isosceles trapezoid. Let F be the foot of the perpendicular from A to $\overline{CE}$, and G be the foot of the perpendicular from B to $\overline{CE}$. Then $EF = CG$. Since $\angle GBC = 30°$, we have

$$CG = \frac{1}{2}(BC) = 1 \text{ and } BG = \frac{\sqrt{3}}{2}(BC) = \sqrt{3}.$$

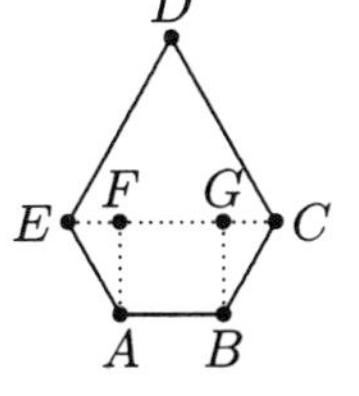

Now

$$CE = CG + GF + FE = 1 + 2 + 1 = 4,$$

so CDE is an equilateral triangle. Thus,

$$[ABCE] = \frac{1}{2}(BG)(AB + CE) = \frac{1}{2}\sqrt{3}\,(2+4) = 3\sqrt{3},$$

$$\text{and } [CDE] = \frac{\sqrt{3}}{4}(CE)^2 = \frac{\sqrt{3}}{4}(16) = 4\sqrt{3}.$$

Therefore,

$$[ABCDE] = [ABCE] + [CDE] = 7\sqrt{3}.$$

OR

Draw $\overline{HI}$ where H is the midpoint of $\overline{ED}$ and I is the midpoint of $\overline{CD}$. Then $ABCIHE$ is a regular hexagon and $\triangle HDI$ is congruent to any of the six equilateral triangles of side 2 that make up $ABCIHE$. Thus, the area of $ABCDE$ is the sum of the areas of 7 equilateral triangles of side 2, so it is $7\left(2^2\frac{\sqrt{3}}{4}\right) = 7\sqrt{3}$.

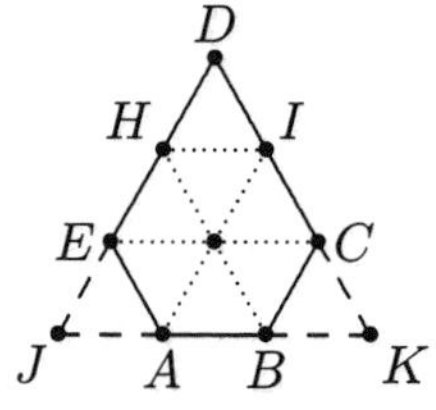

Note. A glance at the diagram for the previous solution shows that the area of $ABCDE$ is tiled by 7 of the 9 congruent equilateral triangles that tile equilateral triangle DJK. Since $JK = 3 \cdot 2 = 6$, the area of $ABCDE$ can by computed as

$$\frac{7}{9}\left(6^2\frac{\sqrt{3}}{4}\right) = 7\sqrt{3}.$$

OR

Extend $\overline{EA}$ and $\overline{CB}$ to meet at P. Since $\angle ABP = \angle BAP = 60°$, $\triangle ABP$ is equilateral as is $\triangle ECP$, and since $EC = CP = CB + BP = 4 = CD = DE$, $\triangle ECD$ is also equilateral. Now,

$$\begin{aligned}[ABCDE] &= ([ECD] + [ECP]) - [ABP] \\ &= 2\left(\frac{\sqrt{3}}{4}4^2\right) - \frac{\sqrt{3}}{4}2^2 = 7\sqrt{3}.\end{aligned}$$

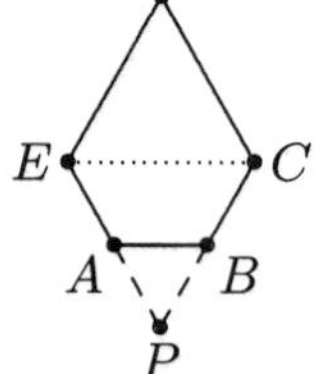

OR

Construct P, and note that $\triangle PAB$ and $\triangle PEC$ are both equilateral, as above. Thus, $\angle AEC = 60°$. Because $\triangle ABC$ is isosceles, $\angle BAC = 30°$, so $\angle CAE = 90°$. Therefore, $\triangle CEA$ is a $30°$-$60°$-$90°$ triangle, so $CE = 4$, $AC = 2\sqrt{3}$ and the altitude from B to $\overline{AC}$ has length 1. Finally, $\triangle CDE$ is equilateral since $CE = 4 = CD = DE$. Hence,

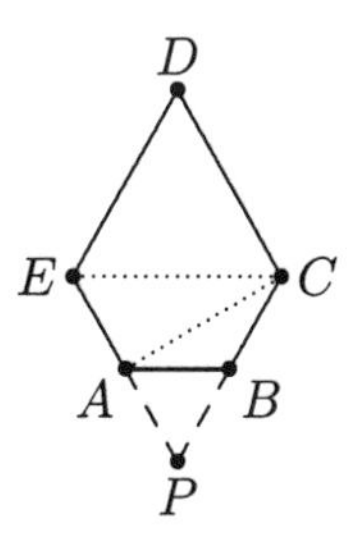

$$\begin{aligned}[ABCDE] &= [ABC] + [CEA] + [CDE] \\ &= \frac{1}{2} \cdot 2\sqrt{3} \cdot 1 + \frac{1}{2} \cdot 2 \cdot 2\sqrt{3} + \frac{\sqrt{3}}{4} \cdot 4^2 = 7\sqrt{3}.\end{aligned}$$

15. **(D)** Since the degree measure of an interior angle of a regular n-sided polygon is

$$\frac{(n-2)180}{n} = 180 - \frac{360}{n},$$

it follows that n must be a divisor of 360. Since $360 = 2^3 3^2 5^1$, its divisors are of the form

$$n = 2^a 3^b 5^c \text{ with } a, b = 0, 1, 2 \text{ or } 3; \ b = 0, 1 \text{ or } 2; \text{ and } c = 0 \text{ or } 1.$$

Hence, there are $(4)(3)(2) = 24$ divisors of 360. Since $n \geq 3$, we exclude the divisors 1 and 2, so there are 22 possible values of n.

OR

After discovering that n must be a divisor of 360, one can simply list the divisors of 360 that are 3 or larger,

$$360, 180, 120, 90, 72, 60, 45, 40, 36, 30, 24, 20,$$

$$3, \ 4, \ 5, \ 6, \ 8, \ 9, 10, 12, 15, 18,$$

and count them, to discover that n can take on 22 possible values.

16. **(D)** The last occurrence of the nth positive integer is in position number

$$1 + 2 + \cdots + n = \frac{n(n+1)}{2}.$$

To approximate our n, consider

$$\frac{n^2}{2} \approx 2000, \quad n \approx \sqrt{4000} = 10\sqrt{40}.$$

Consequently, test $n = 62, 63$:

$$\frac{62(63)}{2} = 1953 \quad \text{and} \quad \frac{63(64)}{2} = 1953 + 63 = 2016.$$

This shows that the last occurrence of 62 is at the 1953rd term of the sequence, and the 2016th term is the last occurrence of 63. Therefore, the 1993rd term is 63, which leaves a remainder of 3 when divided by 5.

17. **(A)** Let O be the center of the clock. Label the triangle from 12 o'clock to 1 o'clock AOB, the quadrilateral from 1 o'clock to 2 o'clock $OBCD$, and the 3 o'clock position E, as indicated in the figure. Then $\triangle AOB \cong \triangle EOD$. Let $AB = 1$. Since $\angle AOB = 30°$, it follows that $OA = \sqrt{3}$, $[OACE] = 3$ and $[AOB] = \sqrt{3}/2$. Hence

$$\begin{aligned}\frac{q}{t} = \frac{[OBCD]}{[AOB]} &= \frac{[OACE] - 2[AOB]}{[AOB]} \\ &= \frac{[OACE]}{[AOB]} - 2 \\ &= \frac{3}{\sqrt{3}/2} - 2 \\ &= 2\sqrt{3} - 2.\end{aligned}$$

A B C
D
E

OR

Note that $\angle AOB = \angle BOD = 30°$. Let $AO = 1$. Therefore $AB = 1/\sqrt{3}$ and $[ABO] = 1/\left(2\sqrt{3}\right)$. Draw $\overline{OC}$. Compute the area of $\triangle OBC$ using $BC = 1 - \left(1/\sqrt{3}\right)$ as the base and $AO = 1$ as the altitude. Then

$$[OBC] = \frac{1}{2} \cdot \left(1 - \frac{1}{\sqrt{3}}\right) \cdot 1 = \frac{3 - \sqrt{3}}{6}.$$

Thus, $\dfrac{q}{t} = \dfrac{[OBCD]}{[AOB]} = \dfrac{2 \cdot \frac{3-\sqrt{3}}{6}}{\frac{1}{2\sqrt{3}}} = 2\sqrt{3} - 2.$

A B C
D
O

OR

Let F be on $\overline{OD}$ so $\overline{BF} \perp \overline{OD}$, and extend $\overline{AC}$ and $\overline{OD}$ to intersect at G. Then

$$\triangle GCD \sim \triangle GFB \cong \triangle OFB \cong \triangle OAB.$$

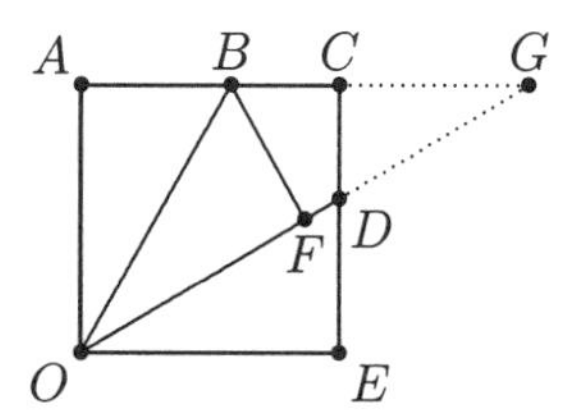

Since

$$GC = GA - CA = GA - OA = OA(\tan 60^\circ - 1) = OA(\sqrt{3} - 1),$$

it follows that $[GCD] = (\sqrt{3}-1)^2[OAB]$. Hence,

$$\begin{aligned}\frac{q}{t} &= \frac{[OBCD]}{[OAB]} = \frac{[OBG] - [GCD]}{[OAB]} \\ &= \frac{2[OAB] - [GCD]}{[OAB]} \\ &= 2 - \frac{[GCD]}{[OAB]} = 2 - (\sqrt{3}-1)^2 = 2\sqrt{3} - 2.\end{aligned}$$

OR

Let $OA = AC = \sqrt{3}$. Then $AB = 1$. Since $\overline{OA}$ is an altitude of both triangles OBC and OAB, it follows that

$$\frac{[OBC]}{[OAB]} = \frac{BC}{AB} = \frac{AC - AB}{AB} = \frac{\sqrt{3}-1}{1}$$

and thus $\dfrac{q}{t} = \dfrac{[OBCD]}{[OAB]} = \dfrac{2[OBC]}{[OAB]} = 2\left(\sqrt{3}-1\right)$.

OR

Note that quadrilateral $OBCD$ is a kite. Let $OA = AC = 1$, so $OC = \sqrt{2}$ and $AB = 1/\sqrt{3}$. Since $\overline{BD}$ is the hypotenuse of the isosceles right triangle BCD,

$$BD = \sqrt{2}BC = \sqrt{2}\left(1 - \frac{1}{\sqrt{3}}\right) = \sqrt{\frac{2}{3}}\left(\sqrt{3}-1\right).$$

Since the area of a kite-shaped region is half the product of its diagonals and the area of a right triangle is half the product of the sides adjacent to the right angle, we have

$$\frac{q}{t} = \frac{(1/2)(OC)(BD)}{(1/2)(OA)(AB)} = \frac{\sqrt{2}\cdot\sqrt{2/3}\left(\sqrt{3}-1\right)}{1\cdot(1/\sqrt{3})} = 2\left(\sqrt{3}-1\right).$$

18. **(E)** Al follows a 4-day cycle, and Barb follows a 10-day cycle. The least common multiple of 4 and 10 is 20. Together they follow a 20-day cycle, and there are 50 such cycles in 1000 days. Let us number the days in each cycle as 1, 2, . . ., 20. Al rests on days 4, 8, 12, 16 and 20. Barb rests on days 8, 9, 10, 18, 19 and 20. They both rest on days 8 and 20, which is 2 days in each cycle. Thus they have $2 \times 50 = 100$ common rest-days during their first 1000 days.

OR

Make a table:

$$\begin{array}{llllllllllll} \mathbf{A}: & www & r & www & \boxed{r} & www & r & www & r & www & \boxed{r} & \cdots \\ \mathbf{B}: & www & w & www & \boxed{r} & rrw & w & www & w & wrr & \boxed{r} & \end{array}$$

Observe that this pattern of 20 days repeats. Since there are 50 such cycles in 1000 days, it follows that the number of common rest-days is $50 \times 2 = 100$.

19. **(D)** Since m and n must both be positive, it follows that $n > 2$ and $m > 4$. Because

$$\frac{4}{m} + \frac{2}{n} = 1 \iff (m-4)(n-2) = 8,$$

we need to find all ways of writing 8 as a product of positive integers. The four ways,

$$1 \cdot 8, \quad 2 \cdot 4, \quad 4 \cdot 2 \text{ and } 8 \cdot 1,$$

correspond to the four solutions

$$(m, n) = (5, 10), \quad (6, 6), \quad (8, 4) \text{ and } (12, 3).$$

OR

Sketch the graph of the hyperbola

$$\frac{4}{x} + \frac{2}{y} = 1 \qquad \text{or} \qquad y = \frac{2x}{x-4}.$$

For x and y to both be positive, we must have $x > 4$ and $y > 2$, so $m \geq 5$ and $n \geq 3$. Observing that $(m, n) = (5, 10), (12, 3)$ are the solutions closest to the asymptotes, we note that $3 \leq n \leq 10$ and test these n to determine which correspond to integers $x = m$ to find the only four possible solutions, $(5, 10)$, $(6, 6)$, $(8, 4)$ and $(12, 3)$.

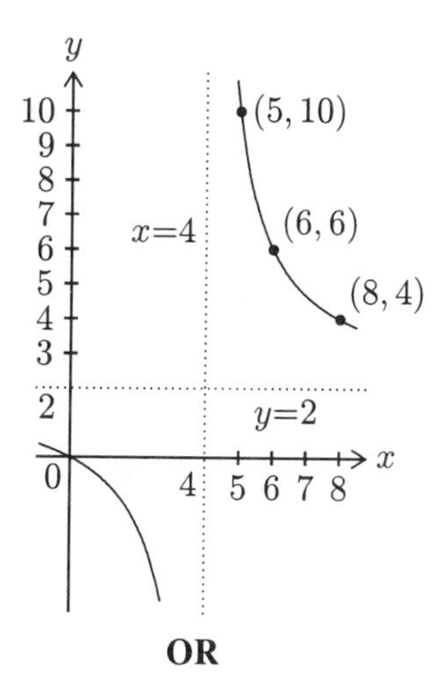

OR

Note that for positive m and n,

$$m > 8 \text{ and } n > 4 \Rightarrow \frac{4}{m} + \frac{2}{n} < 1$$
$$\text{and} \quad m \leq 4 \text{ or } n \leq 2 \Rightarrow \frac{4}{m} + \frac{2}{n} > 1.$$

Divide into cases depending on the relative size of m and 8:

Case 1, $m > 8$: Then $2 < n < 4$, so we test $n = 3$ and find that $(m, n) = (12, 3)$ is a solution

$$\frac{4}{12} + \frac{2}{3} = \frac{1}{3} + \frac{2}{3} = 1.$$

Case 2, $m = 8$: Note that $(m, n) = (8, 4)$ is a solution since

$$\frac{4}{8} + \frac{2}{4} = \frac{1}{2} + \frac{1}{2} = 1.$$

Case 3, $m < 8$: Then $4 < m < 8$, so we search for solutions when $m = 5$, 6 and 7 with $n > 4$. Then $(m, n) = (5, 10)$ and $(m, n) = (6, 6)$ are solutions since

$$\frac{4}{5} + \frac{2}{10} = \frac{4}{5} + \frac{1}{5} = 1$$
$$\text{and} \quad \frac{4}{6} + \frac{2}{6} = \frac{2}{3} + \frac{1}{3} = 1.$$

No solutions are obtained from $m = 7$ since

$$\frac{4}{7} + \frac{2}{n} = 1 \iff 4n + 14 = 7n \iff 3n = 14$$

has no integer solutions for n.

20. **(B)** Use the quadratic formula† to obtain

$$z = \frac{3i \pm \sqrt{-9 + 40k}}{20}.$$

When $k < 0$, then $-9 - 40 < 0$ so both values for z will be pure imaginary numbers. Now find values of k that show all other choices are false:

(A) When $k = 1$, then $z = \left(3i \pm \sqrt{31}\right)/\,20$ which is complex but not pure imaginary.

(C) and **(D)** When $k = i$, then

$$-9 + 40k = -9 + 40i = 16 + 40i - 25 = (4 + 5i)^2,$$

so

$$z = \frac{3i \pm (4 + 5i)}{20} = 4 + 8i/20,\ -4 - 2i/20,$$

neither of which is real.

(E) When $k = 0$, then the roots are $z = 6i/20$ and $z = 0$ which is real.

21. **(B)** In an arithmetic sequence with an odd number of terms, the middle term is the average of the terms. Since a_4, a_7, a_{10} form an arithmetic sequence of three terms with sum 17, $a_7 = 17/3$. Since $a_4, a_5, \ldots, a_{14}$ form an arithmetic sequence of 11 terms whose sum is 77, the middle term, $a_9 = 77/11 = 7$. Let d be the common difference for the given arithmetic sequence. Since $a_7 = 17/3$ and $a_9 = 7$ differ by $2d = 4/3$, $d = 2/3$. Since $a_7 = a_1 + 6d$, it follows that

$$a_1 = a_7 - 6d = 17/3 - 6\left(\frac{2}{3}\right) = \frac{5}{3}.$$

Solve

$$13 = a_k = a_1 + (k-1)d, \quad \text{or} \quad \frac{5}{3} + (k-1)\left(\frac{2}{3}\right) = 13$$

to obtain $k = 18$.

OR

† See footnote to **43, 28**.

Compute $a_9 = 7$ and $d = 2/3$ as above. Since

$$a_k = 13 = 7 + 6 = 7 + 9\left(\frac{2}{3}\right) = a_9 + 9d,$$

it follows that $k = 9 + 9 = 18$.

OR

Let $a = a_1$ and let d be the common difference of the arithmetic sequence. Since $a_k = a + (k-1)d$, it follows that

$$a_4 + a_7 + a_{10} = (a + 3d) + (a + 6d) + (a + 9d) = 17$$

$$3a + 18d = 17 \qquad (\alpha)$$

and $a_4 + a_5 + a_6 \cdots + a_{14}$

$$= (a + 3d) + (a + 4d) + (a + 5d) + \cdots + (a + 13d) = 77$$

$$11a + 88d = 77 \quad (\beta)$$

Solving (α) and (β) simultaneously, we obtain $a = 5/3$ and $d = 2/3$, so

$$13 = a_k = \frac{5}{3} + (k-1)\frac{2}{3} = 1 + \frac{2}{3}k.$$

Therefore, $k = 18$.

22. **(C)** Intuitively, we note that the center cube in the first layer is counted most often and should be assigned the number 1 and those in the corners are used least and should be assigned 8, 9, and 10. For example, to arrive at the correct answer, assign the numbers to the bottom layer in this pattern:

$$\begin{array}{ccccccc} & & & 8 & & & \\ & & 2 & & 3 & & \\ & 7 & & 1 & & 4 & \\ 9 & & 6 & & 5 & & 10 \end{array}$$

More formally, suppose the assignment of the numbers to the bottom layer is:

$$\begin{array}{ccccccc} & & & v_1 & & & \\ & & e_1 & & e_2 & & \\ & e_3 & & c & & e_4 & \\ v_2 & & e_5 & & e_6 & & v_3 \end{array}$$

The arrangement of numbers in the second layer is:

$$(e_1+e_2+v_1)$$

$$(c+e_1+e_3) \qquad (c+e_2+e_4)$$

$$(e_3+e_5+v_2) \qquad (c+e_5+e_6) \qquad (e_4+e_6+v_3)$$

The arrangement of numbers in the third layer is

$$(2c+2e_1+2e_2+e_3+e_4+v_1)$$

$$(2c+e_1+2e_3+2e_5+e_6+v_2) \qquad (2c+e_2+2e_4+e_5+2e_6+v_3)$$

So

$$t = 6c + 3(e_1+e_2+e_3+e_4+e_5+e_6) + (v_1+v_2+v_3)$$

is the number assigned to the top block. Thus, the value of t is minimized when $c = 1$,

$$\{e_1, e_2, e_3, e_4, e_5, e_6\} = \{2, 3, 4, 5, 6, 7\}$$

and $\{v_1, v_2, v_3\} = \{8, 9, 10\}$. Hence, the minimum value of t is

$$6(1) + 3(2 + 3 + 4 + 5 + 6 + 7) + (8 + 9 + 10) = 114.$$

Comment. The sum of all the numbers assigned to the bottom layer is 55. We could therefore compute

$$\begin{aligned} t &= 55 + 2(e_1+e_2+e_3+e_4+e_5+e_6) + 5c \\ &= 55 + 2(27) + 5 = 114. \end{aligned}$$

23. **(B)** The center of the circle is not X since $2\angle BAC \neq \angle BXC$. Thus $\overline{AD}$ bisects $\angle BXC$ and $\angle BAC$. Since $\angle ABD$ is inscribed in a semicircle, $\angle ABD = 90°$, and thus

$$AB = AD \cos \angle BAD = 1 \cdot \cos\left(\frac{1}{2}\angle BAC\right) = \cos 6°.$$

Also,

$$\angle AXB = 180° - \angle DXB = 180° - \frac{36°}{2} = 162°.$$

Since the sum of the angles in $\triangle AXB$ is $180°$,

$$\angle ABX = 180° - (162° + 6°) = 12°.$$

By the *Law of Sines*,

$$\frac{AB}{\sin \angle AXB} = \frac{AX}{\sin \angle ABX} \quad \text{so} \quad \frac{\cos 6^\circ}{\sin 162^\circ} = \frac{AX}{\sin 12^\circ}.$$

Since $\sin 162^\circ = \sin 18^\circ$, we have

$$AX = \frac{\cos 6^\circ \sin 12^\circ}{\sin 18^\circ} = \cos 6^\circ \sin 12^\circ \csc 18^\circ.$$

Note. If X were the center of the circle, then $\overline{BX}$ and $\overline{CX}$ could be any radii. Therefore, to establish $\overline{AD}$ as the bisector of $\angle BAC$ and $\angle BXC$, it is necessary first to establish that X is not the center of the circle.

24. **(E)** Compute the probability of drawing the 3 shiny pennies in 3 or 4 draws, and subtract this from 1. When all 7 pennies are drawn, there are $\binom{7}{3}$ positions the shiny pennies can occupy. Similarly, if among the first 4 pennies drawn there are 3 shiny pennies, then there are $\binom{4}{3}$ positions for these 3 shiny pennies. Thus, the probability that the 3 shiny pennies are drawn among the first 4 is

$$\frac{\binom{4}{3}}{\binom{7}{3}} = \frac{\frac{4!}{3!1!}}{\frac{7!}{3!4!}} = \frac{4}{35}.$$

Thus the requested probability is $1 - (4/35) = 31/35$, whose numerator and denominator sum to 66.

OR

The probability that the three shiny pennies come up in 4 or fewer draws is the probability of getting 3 shiny pennies among the first 4 drawn. There are $\binom{3}{3}$ ways of selecting the shiny pennies, and $\binom{4}{1}$ ways to select the dull penny. Since there are $\binom{7}{4}$ ways to choose the first 4 pennies, the probability that the 3 shiny pennies will be among the first 4 is

$$\frac{\binom{3}{3}\binom{4}{1}}{\binom{7}{4}} = \frac{4}{35}.$$

The complement of this is

$$1 - \frac{4}{35} = \frac{31}{35},$$

whose numerator and denominator sum to 66.

OR

The paths from the origin O to $P = (4, 3)$ in the (d, s)-plane [See figure.] using 7 unit steps, each step being either in the direction of the positive d-axis (**D**) or in the direction of the positive s-axis (**S**), can be represented as a sequence of 7 **D**'s and **S**'s, exactly 3 of which are **S**'s. The set of all such paths corresponds to the set of all possible orders in which all 7 pennies, 4 dull and 3 shiny, can be drawn. The set of all such paths which avoid the dotted portion of the grid corresponds to the set of orders for drawing the pennies in which it takes more than four draws to obtain the third shiny penny. The paths from O to P which avoid the dotted portion of the grid must use one of $\overline{(2,2)(2,3)}$, $\overline{(3,2)(3,3)}$ or $\overline{(4,2)(4,3)}$ as their last step upward.

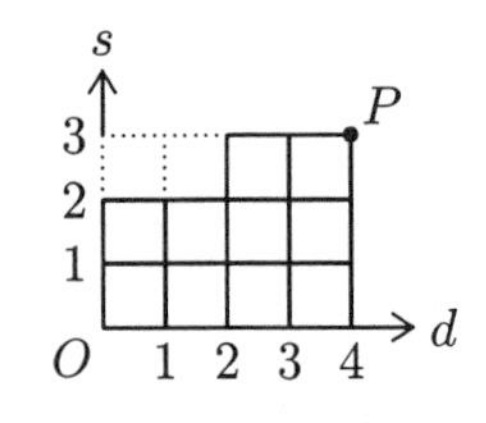

The total number of paths from O to P is the number of ways of choosing places for 3 **S**'s in □□□□□□□ (and filling the remaining 4 places with **D**'s), and that is $\binom{7}{3} = 35$ ways. The number of these which use $\overline{(2,2)(2,3)}$ as their last upward path is the number of ways of placing two **S**'s in □□□□, which is $\binom{4}{2} = 6$. Similar analysis shows that the numbers of paths using $\overline{(3,2)(3,3)}$ and $\overline{(4,2)(4,3)}$ as their last upward step are $\binom{5}{2} = 10$ and $\binom{6}{2} = 15$, respectively. Thus, the desired probability is $(6+10+15)/35 = 31/35$, and $a + b = 66$.

Comment. Some mathematicians and problem-solvers consider paths on a grid, as used in the last solution, to be the most natural way to introduce combinations and the binomial coefficients.

25. **(E)** The figure shows the original configuration with the vertex of the $120°$ angle labeled O and three additional line segments drawn from P to S: AP, QP, RP. Point A is chosen so that $\angle APO = 60°$. Thus $\triangle APO$ is equilateral. Point Q is **any** point on segment $\overline{AO}$, and R is on the other side of O with $AQ = OR$. Since $PO = PA$, $\angle POR \cong \angle PAQ$, and $OR = AQ$, we have $\triangle POR \cong \triangle PAQ$. Thus, $\angle OPR \cong \angle APQ$, and hence $\angle QPR \cong \angle APO = 60°$. Since $QP = RP$ and $\angle QPR = 60°$, $\triangle QPR$ is also equilateral. Since Q was an arbitrary point on segment $\overline{AO}$, it follows that there are an infinite number of the required equilateral triangles.

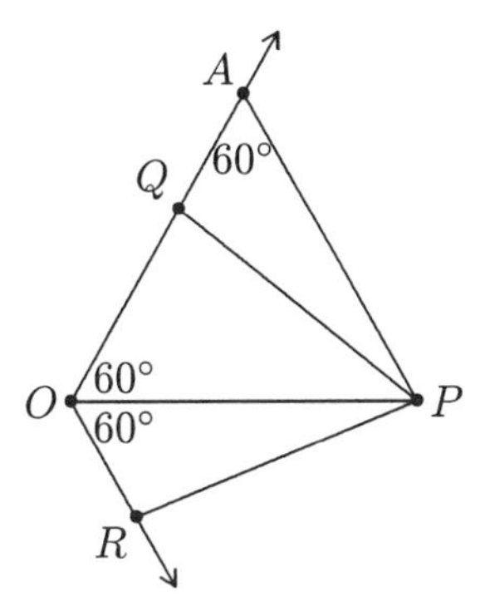

Comment. The fact that the given 120° angle is the supplement of the 60° interior angle of an equilateral triangle very strongly motivates the investigation of connections between this problem and cyclic quadrilaterals. This connect is pursued in the next two solutions.

OR

As above, let O be the vertex of the 120° angle, choose A on the one ray so $\angle APO = 60^\circ$ and choose **any** point Q on segment $\overline{AO}$. Now choose R on the other ray with $\angle QPR = 60^\circ$. Then $PQOR$ is a cyclic quadrilateral since angles QPR and QOR are supplementary. The minor arc PR of the circumscribed circle is subtended by both $\angle POR$ and $\angle PQR$, so $\angle PQR = \angle POR = 60^\circ$. Similarly, $\angle PRQ = \angle POQ = 60^\circ$. Since $\angle QPR = 60^\circ$, $\triangle QPR$ is equiangular and therefore equilateral. Since the point Q on segment $\overline{AO}$ was arbitrary, there are an infinite number of equilateral triangles PQR.

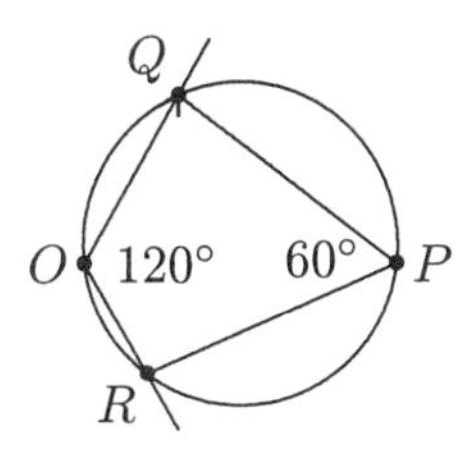

Comment. The following solution also uses a cyclic quadrilateral, but the circle is constructed first. Whenever a problem involves four or more points, it is not a bad idea to check to see if a cyclic quadrilateral might be of use.

OR

Let O be the vertex of the 120° angle, choose **any** circle which has $\overline{OP}$ as a chord, and let Q and R be the other intersections of the set S with the circle. Since $OQPR$ is inscribed in a circle, $\angle QPR = 60^\circ$ because it is the supplement of $\angle QOR$. We have $\angle PRQ = \angle POQ$ since both intercept the same arc, and similarly, $\angle RQP = \angle ROP$. Triangle PQR is equilateral because $\angle QPR = 60^\circ = \angle PRQ = \angle RQP$. Since the circle with $\overline{OP}$ as chord was arbitrary, there are an infinite number of equilateral triangles PQR.

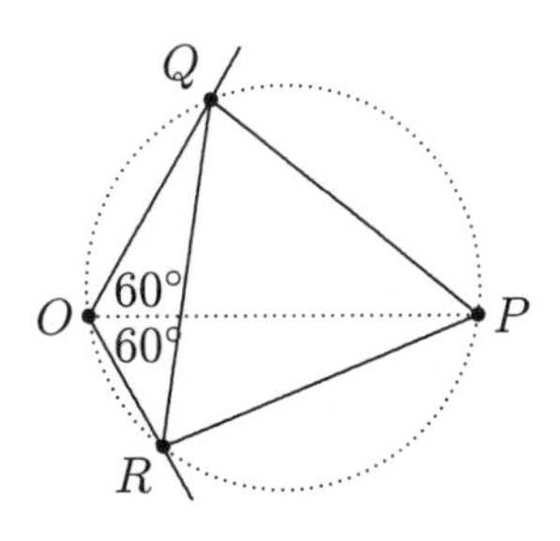

Query. If the rays form an angle different from 120°, would there still be an infinite number of equilateral triangles? If not, how many?

26. **(C)** Completing the squares, we have

$$f(x) = \sqrt{16 - (x-4)^2} - \sqrt{1 - (x-7)^2}.$$

The first term is the formula for the y-coordinate of the upper half of the circle with center at $(4,0)$ and radius 4, and the second term is the formula for the y-coordinate of the upper half-circle with center at $(7,0)$ and radius 1. The graphs of these two semicircles show that $f(x)$ is real-valued only when $6 \le x \le 8$. On that interval, the value of $f(x)$ is the difference in the heights of the semicircles, so we can see that the maximum value of $f(x)$ will be attained when $x = 6$.

Evaluating, we get

$$f(6) = \sqrt{16 - (6-4)^2} = \sqrt{12} = 2\sqrt{3}.$$

OR

Note that

$$f(x) = \sqrt{(8-x)x} - \sqrt{(8-x)(x-6)}$$

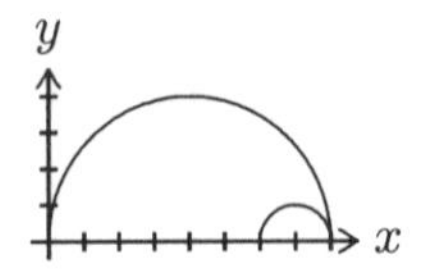

is a real number if and only if $6 \leq x \leq 8$. Factor and multiply by the conjugate:

$$f(x) = \sqrt{8-x}\left(\sqrt{x} - \sqrt{x-6}\right)\left(\frac{\sqrt{x}+\sqrt{x-6}}{\sqrt{x}+\sqrt{x-6}}\right)$$
$$= \frac{6\sqrt{8-x}}{\sqrt{x}+\sqrt{x-6}}.$$

For all x such that $6 \leq x \leq 8$, the numerator of this last expression is maximized and its denominator is minimized when $x = 6$. Hence, the maximum value of this fraction is

$$f(6) = \frac{6\sqrt{2}}{\sqrt{6}} = 2\sqrt{3}.$$

Comment. This is a problem which quite possibly might be more challenging to a participant with a rudimentary knowledge of calculus, especially if the domain of the function is not considered. Calculus is not a prerequisite for the AHSME. The AHSME committee will not consider any problem which can be solved more easily using calculus than without. Consequently, experienced AHSME participants know that good approaches to problems involving extremals usually involve either completing squares or an application of the Arithmetic-Geometric Mean Inequality.

27. **(B)** When the circle is closest to A with its center P at A', let its points of tangency to $\overline{AB}$ and $\overline{AC}$ be D and E, respectively. The path parallel to $\overline{AB}$ is shorter than $\overline{AB}$ by AD plus the length of a similar segment at the other end. Now $AD = AE = \cot(A/2)$. Similar reasoning at the other vertices shows that the length L of the path of P is

$$L = AB + BC + CA - 2\cot\frac{A}{2} - 2\cot\frac{B}{2} - 2\cot\frac{C}{2}.$$

Note that

$$\cot\frac{A}{2} = \left(\frac{\cos(A/2)}{\sin(A/2)}\right)\left(\frac{2\cos(A/2)}{2\cos(A/2)}\right) = \frac{2\cos^2(A/2)}{2\sin(A/2)\cos(A/2)}$$

$$= \frac{1+\cos^2(A/2)-\sin^2(A/2)}{\sin A}$$
$$= \frac{1+\cos A}{\sin A} = \frac{1+(4/5)}{3/5} = 3.$$

Similarly
$$\cot\frac{C}{2} = \frac{1+\cos C}{\sin C} = \frac{1+(3/5)}{4/5} = 2.$$

Of course, $\cot(B/2) = \cot 45^\circ = 1$. The length of the path is
$$L = 8+6+10-2(3)-2(1)-2(2) = 12.$$

Note. One could also obtain $\cot(A/2)$ using
$$\cot\frac{A}{2} = \sqrt{\frac{1+\cos A}{1-\cos A}} = \sqrt{\frac{1+\frac{4}{5}}{1-\frac{4}{5}}} = \sqrt{\frac{9/5}{1/5}} = 3,$$
and similarly
$$\cot\frac{C}{2} = \sqrt{\frac{1+\cos C}{1-\cos C}} = \sqrt{\frac{1+\frac{3}{5}}{1-\frac{3}{5}}} = \sqrt{\frac{8/5}{2/5}} = 2.$$

OR

The locus of P is a triangle $A'B'C'$ similar to triangle ABC since the sides of $\triangle A'B'C'$ are parallel to the respective sides of $\triangle ABC$. We compute only $A'B'$:
$$A'B' = AB - \cot\frac{A}{2} - \cot\frac{B}{2} = 8-3-1 = 4 = \frac{1}{2}AB.$$
Thus the linear dimensions of $\triangle A'B'C'$ are half those of $\triangle ABC$. Consequently, the perimeter of $\triangle A'B'C'$ is half that of $\triangle ABC$, $(8+6+10)/2 = 12$.

OR

The diagram with the first solution shows that the line $\overline{AA'}$, bisects $\angle A$. It will also bisect $\angle A'$. Similarly, lines $\overline{BB'}$ and $\overline{CC'}$ are bisectors of $\angle B$ and $\angle B'$, and of $\angle C$ and $\angle C'$, respectively. Therefore $\triangle ABC$ and $\triangle A'B'C'$ have the same incenter because they have the same angle bisectors. The inradius of $\triangle A'B'C'$ will be one less than that of $\triangle ABC$. The inradius of a right triangle with legs a and b and hypotenuse c is $r = ab/(a+b+c)$ because the area of the triangle can be expressed either as $ab/2$ or as $r(a+b+c)/2$. Thus, the inradius of $\triangle ABC$ is $(8)(6)/(8+6+10) = 2$, and therefore the inradius of

$\triangle A'B'C'$ is $2-1=1$. Since the ratio of the inradii is the same as the ratio of the perimeters, the perimeter of $\triangle A'B'C'$ is $(8+6+10)/2 = 12$.

Note. One can also obtain the radius r of the inscribed circle of a right triangle with hypotenuse c using

$$r = \frac{a+b-c}{2}.$$

This follows from the fact that the lengths of the tangents to the circle from the right angle have length r. Using this formula, not surprisingly, we also have

$$r = \frac{6+8-10}{2} = 2.$$

OR

Extend $\overline{C'A'}$ to point E so $\overline{AE} \perp \overline{EA'}$, and let $\overline{EA'}$ intersect $\overline{AB}$ at D. Let F be the point on $\overline{AB}$ such that $\overline{A'F} \perp \overline{AB}$. Then, since

$$\angle ADE = \angle A'DF = \angle CAB,$$

we have

$$\triangle AED \sim \triangle A'FD \sim \triangle CBA.$$

Since $AE = A'F = 1$, it follows that

$$AF = AD + DF = \frac{10}{6} + \frac{8}{6} = 3,$$

so

$$\begin{aligned} A'B' &= AB - AF - 1 \\ &= 8 - 3 - 1 = 4 = \frac{AB}{2}. \end{aligned}$$

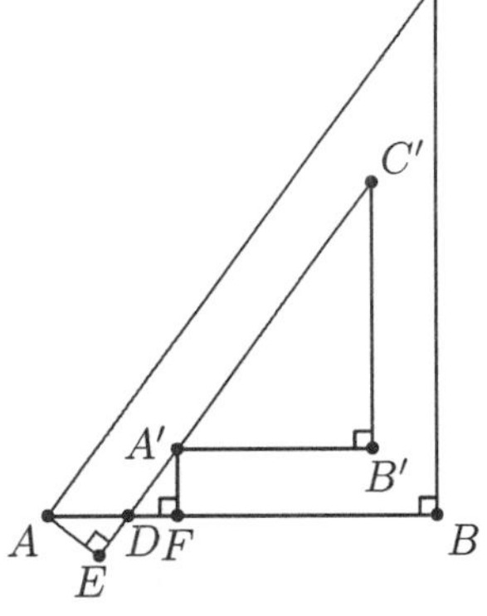

Thus, the linear dimensions of $\triangle A'B'C'$ are half those of $\triangle ABC$, so the perimeter of $\triangle A'B'C'$ is $(8+6+10)/2 = 12$.

OR

The locus, $\triangle DEF$, of the center of the rolling circle is similar to $\triangle ABC$, so we label its sides $3x$, $4x$ and $5x$, for some $x > 0$. The area of $\triangle ABC$ is

$$\mathcal{A} = [ABC] = (AB)(BC)/2 = 24.$$

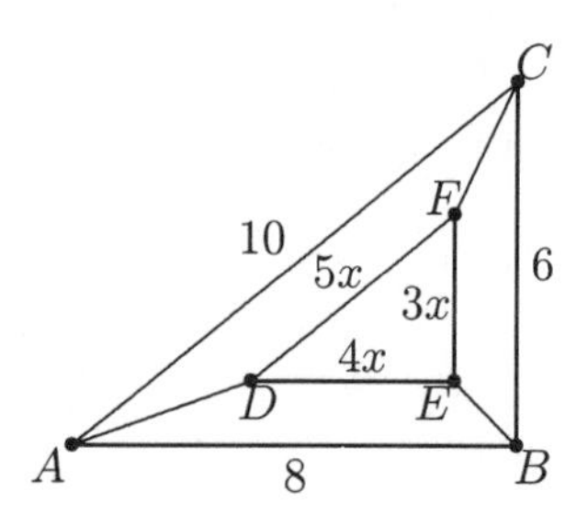

Partition $\triangle ABC$ into three trapezoids of altitude 1 together with $\triangle DEF$, and compute the area $\mathcal{A}$ of $\triangle ABC$ in terms of x:

$$\begin{aligned}\mathcal{A} &= [DABE] + [EBCF] + [FCAD] + [DEF]\\ 24 &= \frac{1}{2}(1)(4x+8) + \frac{1}{2}(1)(3x+6) + \frac{1}{2}(1)(5x+10) + \frac{12x^2}{2}\\ 24 &= 6x^2 + 6x + 12.\end{aligned}$$

Solve

$$6x^2 + 6x - 12 = 0$$

for the positive root, $x = 1$, to find that the perimeter of $\triangle DEF$ is $3x + 4x + 5x = 12x = 12$.

OR

Put the triangle in the Cartesian plane with $A = (8,0)$, $B = (0,0)$ and $C = (0,6)$. When the circle rolls along $\overline{BA}$, P will move along the line $y = 1$. When the circle rolls along $\overline{CB}$, P will move along the line $x = 1$. When the circle rolls along $\overline{AC}$, P will move along a line ℓ parallel to $\overline{AC}$ but at a distance 1 from $\overline{AC}$. To find the y-intercept $(0,b)$ of line ℓ, note that $\triangle BAC$ is similar to the triangle with vertices $(0,4)$, $(1,4)$ and $(0,b)$. Since side $\overline{(0,4)(1,4)}$ is of length one, and $BA = 8$, it follows that the ratio of all the linear dimensions is 8. Therefore, $8(b-4) = BC = 6$, so $b = 19/4$. Hence ℓ has the equation

$$y = -\frac{3}{4}x + \frac{19}{4}.$$

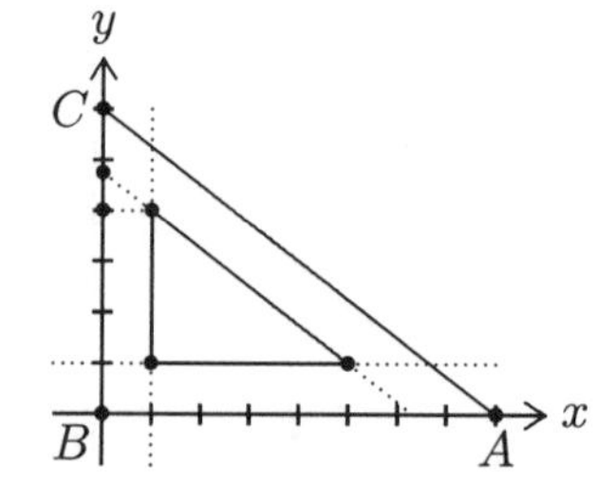

The pairwise intersection points of this line and the two lines $x = 1$ and $y = 1$ are the vertices of the triangular path traced by P. These points are $(1,1)$, $(1,4)$, $(5,1)$. It now follows that P travels a distance of

$$(4-1)+\sqrt{(5-1)^2+(1-4)^2}+(5-1)=12.$$

Query. The following note gives a geometric model for the difference between the perimeters of the two triangles. That model obviously does not depend on the size of the angles, so it can be generalized to any triangle. What is the model for the difference between the perimeter of an isosceles trapezoid and the locus of the center of a circle rolling around inside it?

Note. The perimeter of $\triangle A'B'C'$, generated by the center of the rolling circle, is the perimeter of $\triangle ABC$ minus the perimeter of a triangle $A''B''C''$ that is similar to $\triangle ABC$ and whose incircle is congruent to the rolling circle. To see this, let L_1, L_2 on $\overline{AB}$, M_1, M_2 on $\overline{BC}$ and N_1, N_2 on $\overline{CA}$ be the points where the circle is simultaneously tangent to two sides of $\triangle ABC$. [See top figure.] Note that quadrangles $A'N_2AL_1$, $B'L_2BM_1$ and $C'M_2CN_1$ can be re-assembled into $\triangle A''B''C''$ with the circle inscribed. [See bottom figure.] Then

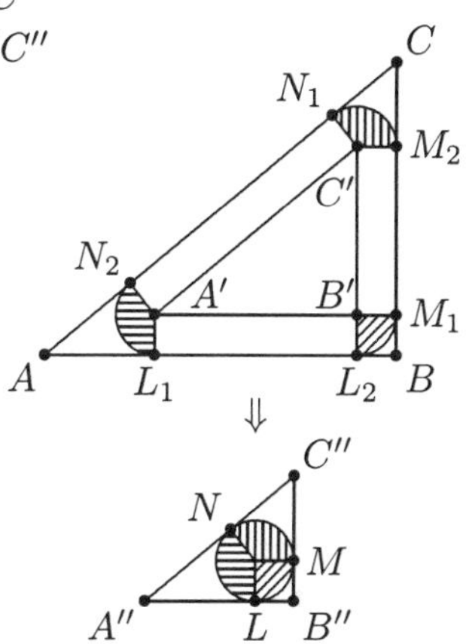

$$\begin{aligned}A'B'+B'C'+C'A' &= L_1L_2+M_1M_2+N_1N_2\\ &= (AB+BC+CA)-\big((AL_1+L_2B)\\ &\quad +(BM_1+M_2C)+(CN_1+N_2A)\big)\\ &= (AB+BC+CA)-(A''B''+B''C''+C''A'').\end{aligned}$$

28. **(D)** There are $\binom{16}{3}=560$ sets of 3 points. We must exclude from our count those sets of three points that are collinear. There are 4 vertical and 4 horizontal lines of four points each. These 8 lines contain $8\binom{4}{3}=32$ sets of 3 collinear points. [See top figure.] There are

$$\binom{3}{3}+\binom{4}{3}+\binom{3}{3}=6$$

sets of collinear points on lines of slope $+1$ [See bottom figure.], and a similar number on lines of slope -1. Since there are no other sets of 3 collinear points, the number of triangles is $560 - 32 - 12 = 516$.

Challenge. Find the number of triangles with vertices in a 5×5, a 6×6, or even an $n \times n$ array of lattice points.

29. **(B)** Let a, b and c, with $a \le b \le c$, be the lengths of the edges of the box; and let p, q and r, with $p \le q \le r$, be the lengths of its external diagonals. (The diagram shows three faces adjacent at a vertex "flattened out" as in the net of a solid.) The Pythagorean Theorem implies that

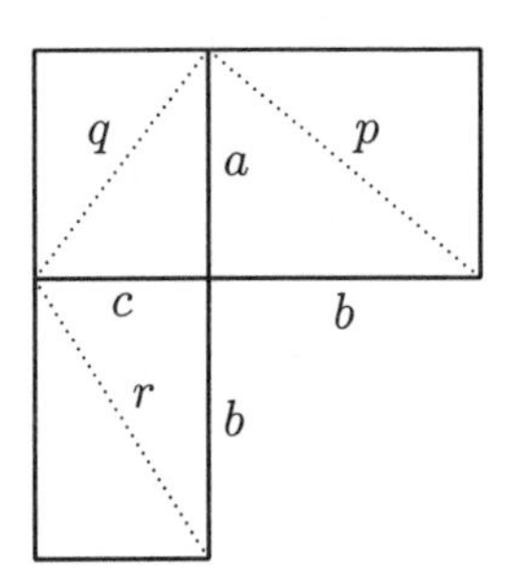

$$\begin{aligned} p^2 &= a^2 + b^2 \\ q^2 &= a^2 + c^2 \\ r^2 &= b^2 + c^2, \end{aligned}$$

so $r^2 = p^2 + q^2 - 2a^2 < p^2 + q^2$

is a necessary condition for a set $\{p, q, r\}$ to represent the lengths of the diagonals. Only choice **(B)** fails this test:

(A) $6^2 = 36 < 41 = 4^2 + 5^2$; **(B)** $7^2 = 49 \not< 41 = 4^2 + 5^2$;
(C) $7^2 = 49 < 52 = 4^2 + 6^2$; **(D)** $7^2 = 49 < 61 = 5^2 + 6^2$;
(E) $8^2 = 64 < 74 = 5^2 + 7^2$.

The other four choices *do* correspond to actual prisms because the condition $r^2 < p^2 + q^2$ is also sufficient. To see this, just solve the equations for a, b and c:

$$\begin{aligned} a^2 &= \frac{p^2 + q^2 - r^2}{2}, \\ b^2 &= \frac{p^2 - q^2 + r^2}{2}, \\ c^2 &= \frac{-p^2 + q^2 + r^2}{2}. \end{aligned}$$

OR

The angle θ, formed at a vertex of the parallelepiped by diagonals of two adjacent faces, is less than the 90° dihedral angle formed by the two faces. It follows that a triangle formed by choosing one diagonal in each of three faces must be an acute triangle. Therefore, if p, q and r are the lengths of the face diagonals with $r \geq p, q$, then from the *Law of Cosines* it follows that

$$r^2 = p^2 + q^2 - 2pq\cos\theta < p^2 + q^2$$

since $\cos\theta > 0$.

30. **(D)** When a number $x_0 \in [0, 1)$ is written in base-two, it has the form

$$x_0 = 0.d_1d_2d_3d_4d_5d_6d_7\ldots \qquad \text{(each } d_k = 0 \text{ or } 1.)$$

The algorithm given in the problem simply moves the "binary point" one place to the right and then ignores any digits to the left of the point. That is

$$x_0 = 0.d_1d_2d_3d_4d_5d_6d_7\ldots \quad \Longrightarrow \quad x_1 = 0.d_2d_3d_4d_5d_6d_7d_8\ldots$$

Thus for x_0 to equal x_5 we must have

$$0.d_1d_2d_3d_4d_5d_6d_7\ldots = 0.d_6d_7d_8d_9d_{10}d_{11}d_{12}\ldots$$

This happens if and only if x_0 has a repeating expansion with

$$d_1d_2d_3d_4d_5$$

as the repeating block. There are $2^5 = 32$ such blocks. However, if $d_1 = d_2 = d_3 = d_4 = d_5 = 1$, then $x_0 = 1$. Hence there are $32 - 1 = 31$ values of $x_0 \in [0, 1)$ for which $x_0 = x_5$.

OR

We can restate the given formula as $x_n = 2x_{n-1} - \lfloor 2x_{n-1} \rfloor$, where $\lfloor t \rfloor$ is the largest integer not exceeding t. Since $\lfloor t + k \rfloor = \lfloor t \rfloor + k$ for any integer k, it follows that

$$\begin{aligned}
x_5 &= 2x_4 - \lfloor 2x_4 \rfloor \\
&= 2(2x_3 - \lfloor 2x_3 \rfloor) - \lfloor 2(2x_3 - \lfloor 2x_3 \rfloor) \rfloor \\
&= 4x_3 - 2\lfloor 2x_3 \rfloor - \lfloor 4x_3 \rfloor + 2\lfloor 2x_3 \rfloor \\
&= 4x_3 - \lfloor 4x_3 \rfloor \\
&\ \ \vdots \\
&= 32x_0 - \lfloor 32x_0 \rfloor .
\end{aligned}$$

Consequently, to have $x_5 = x_0$ it is necessary that $31x_0 = \lfloor 32x_0 \rfloor$ be an integer. But $31x_0$ is an integer for some x_0 in the prescribed domain precisely when $x_0 = n/31$ for some $n = 0, 1, \ldots, 30$. Note that for such n, $0 \leq (n/31) < 1$ so

$$\left\lfloor 32\left(\frac{n}{31}\right)\right\rfloor = \left\lfloor \frac{32}{31}n \right\rfloor = \left\lfloor n + \frac{n}{31} \right\rfloor = n.$$

Therefore, when $x_0 = n/31$,

$$\begin{aligned} x_5 &= 32x_0 - \lfloor 32x_0 \rfloor \\ &= 32\left(\frac{n}{31}\right) - \left\lfloor 32\left(\frac{n}{31}\right)\right\rfloor = \frac{32}{31}n - n = \frac{n}{31} = x_0. \end{aligned}$$

OR

Note that

$$x_2 = \begin{cases} 4x_0 & \text{if } 0 \leq 4x_0 < 1 \\ 4x_0 - 1 & \text{if } 0 \leq 4x_0 - 1 < 1 \\ 4x_0 - 2 & \text{if } 0 \leq 4x_0 - 2 < 1 \\ 4x_0 - 3 & \text{if } 0 \leq 4x_0 - 3 < 1 \end{cases}$$

and

$$x_3 = \begin{cases} 8x_0 & \text{if } 0 \leq 8x_0 < 1 \\ 8x_0 - 1 & \text{if } 0 \leq 8x_0 - 1 < 1 \\ & \vdots \\ 8x_0 - 7 & \text{if } 0 \leq 8x_0 - 7 < 1. \end{cases}$$

Continuing in this way, one can show that

$$0 \leq x_5 = 32x_0 - r < 1 \text{ for some } r \in \{0, 1, 2, \ldots, 31\}.$$

With $x_5 = x_0$ we have $x_0 = 32x_0 - r$, or $x_0 = r/31$, for a total of 31 choices for x_0. It can be verified that for each of these 31 values x_0 we have $x_5 = x_0$.

OR

Let

$$f(x) = \begin{cases} 2x & \text{if } 2x < 1 \\ 2x - 1 & \text{if } 2x \geq 1, \end{cases}$$

and note that to obtain the graph $\mathcal{F}$ of $y = f(g(x))$ from the graph $\mathcal{G}$ of $y = g(x)$ one first 'stretches' the plane in the vertical direction by a factor of 2 and then 'lowers' the points at or above $y = 1$ by 1 unit. On the interval $0 \leq x < 1$ we have used solid lines to show $y = x$ in the leftmost sketch, and the result of applying this process first to

$y = x$ to obtain $y = f(x)$, then to $y = f(x)$ to obtain $y = f^2(x)$, and finally to $y = f^2(x)$ to obtain $y = f^3(x)$ in the other sketches.

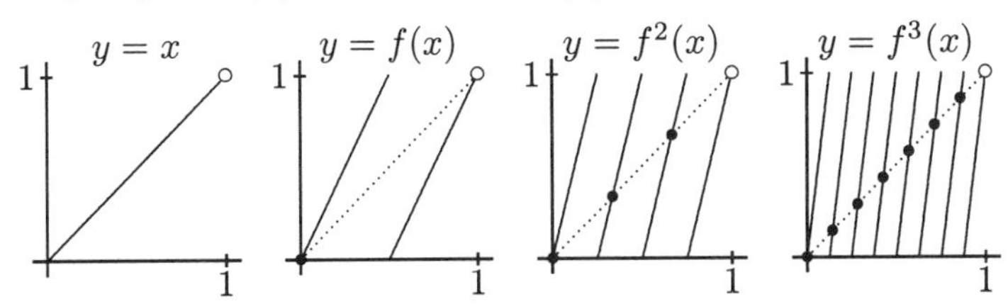

The answer to this problem is the number of intersections of $y = x$ with $y = f^5(x)$ on the interval $0 \leq x < 1$. The dotted lines in the last three sketches indicate $y = x$, and induction can be used to show that the number of intersections of $y = x$ with $y = f^n(x)$ for any $n > 1$ is $2^n - 1$. When $n = 5$, the number of intersections is therefore $2^5 - 1 = 31$.

45 AHSME Solutions

1. **(C)** $4^4 \cdot 9^4 \cdot 4^9 \cdot 9^9 = (4 \cdot 9)^4 \cdot (4 \cdot 9)^9 = (4 \cdot 9)^{4+9} = 36^{13}$.

OR

$$4^4 \cdot 9^4 \cdot 4^9 \cdot 9^9 = (4^{4+9}) \cdot (9^{4+9}) = 4^{13}9^{13} = (4 \cdot 9)^{13} = 36^{13}.$$

2. **(B)** Rectangles of the same height have areas proportional to their bases, a and b. Hence

$$\frac{6}{14} = \frac{a}{b} = \frac{?}{35},$$

so the required area is 15.

OR

We can prove that the product of the areas of the diagonally opposite rectangles is the same: If segment lengths are x, y, u and v as shown, then the product of the areas of the lower left and upper right is $(xv)(yu)$ and the product of the other areas is $(xu)(yv)$. Since $(35)(6) = (14)(?)$, the area of the fourth rectangle is $(35)(6)/14 = 15$.

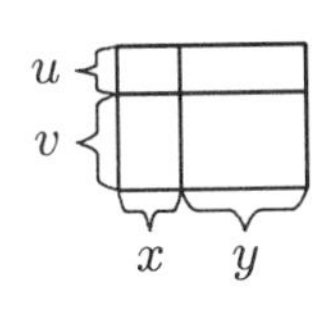

Note. For the given data, we can find the answer by factoring the areas. Since rectangles of the same height have areas proportional to their bases, the unknown area is $r \cdot s = 3 \cdot 5 = 15$. This factoring method will not work for all given data; e.g., if 16 is substituted for 14.

$3 \cdot 2$	$7 \cdot 2$
$r \cdot s$	$7 \cdot 5$

OR

Let a be the height of the rectangles of areas 6 and 14. Then their respective widths are $6/a$ and $14/a$. Since the width of the rectangle of area 35 is $14/a$, its height is $35/(14/a)$, so the area of the fourth rectangle is

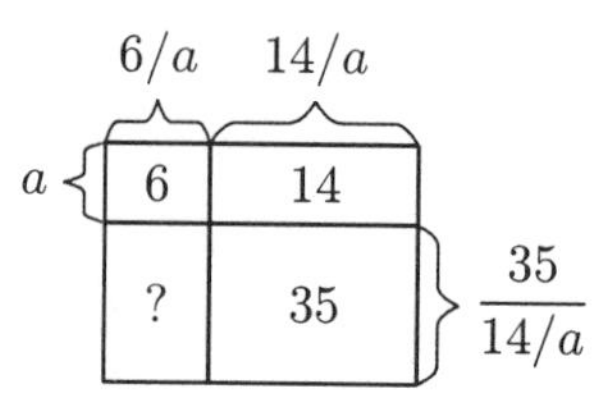

$$\left(\frac{6}{a}\right)\left(\frac{35}{14/a}\right) = 15.$$

3. **(B)** Only one, namely $2x^x$, equals $x^x + x^x$ for all $x > 0$. Use $x = 2$ or $x = 3$ to show that the given expression is not identical to any of the other choices.

4. **(A)** The radius of the circle is $[25 - (-5)]/2 = 15$, and the midpoint of the diameter is $(10, 0)$. Thus an equation of the circle is $(x-10)^2 + y^2 = 15^2$. Let $y = 15$ in this equation to find that $x = 10$.

OR

The circle has center $(10, 0)$, diameter 30, and hence radius 15. Since $(x, 15)$ is 15 units from the given diameter, the radius to $(x, 15)$ must be perpendicular to that diameter. Thus $x = 10$.

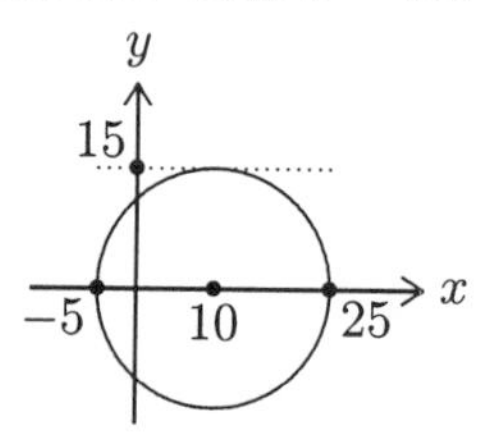

5. **(E)** The original number n satisfies

$$\frac{n}{6} - 14 = 16,$$

so $n/6 = 30$ and $n = 180$. Thus, the answer Pat should have produced is $6(180) + 14 = 1094$.

6. **(A)** Use the rule for generating terms of the sequence beginning with the two terms a, b and working to the right:

$$a,\ b,\ a+b,\ a+2b,\ 2a+3b,\ 3a+5b,\ \ldots$$

to see that

$$2a + 3b = 0$$

$$\text{and} \quad 3a + 5b = 1.$$

Solve these equations simultaneously to find $a = -3$.

OR

Calculate d, c, b and a in that order from the generating rule for the sequence:

$1 = d + 0$, so $\boxed{d = 1}$;
then $0 = c + d = c + 1$, so $\boxed{c = -1}$;
next $1 = d = b + c = b - 1$, so $\boxed{b = 2}$;
finally, $-1 = c = a + b = a + 2$, so $\boxed{a = -3}$.

Challenge. We seek x_{-4} where $x_0 = 0$, $x_1 = 1$ and

$$x_{n+2} = x_{n+1} + x_n.$$

Generalize the previous solution to show that, $|x_j| = |x_{-j}|$ and x_j is negative if and only if j is negative and even.

7. **(E)** Each square has side 10, and thus area 100. The overlap, $\triangle ABG$, has area one-fourth that of square $ABCD$. Thus, the area covered is $100 + 100 - (100/4) = 175$.

OR

Since $\triangle ABG$ is an isosceles right triangle with hypotenuse $AB = 10$, it follows that the length of each of its legs is $5\sqrt{2}$ so its area is $(5\sqrt{2})\,(5\sqrt{2})\,/\,2 = 25$. Thus, the total area covered is $2(100) - 25 = 175$.

Note. It is not necessary that $\overline{GH}$ and $\overline{GF}$ coincide with the diagonals of square $ABCD$. The area of the overlap is constant as long as G is the center of $ABCD$.

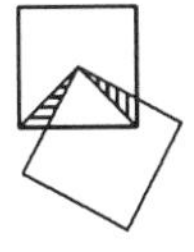

8. **(C)** It is possible to sketch the polygon on a 7×7 square grid. Let the length of a side of the polygon be s. The perimeter of the polygon is $28s$, so $s = 2$. The region bounded by the polygon consists of

$$1 + 3 + 5 + 7 + 5 + 3 + 1 = 25$$

squares, so its area is $25s^2 = 100$.

OR

Instead of counting the 25 squares as above, note that to form the figure from the 7×7 grid, 6 $s \times s$ squares are cut from each corner. Hence the area of the region is $[(7)(7) - 4(6)]s^2 = 100$.

OR

As shown, remove each of the four squares a, b, c, d furthest from the center of the polygon, and use them to fill in the remaining four concave sections, A, B, C, D. The area of the region bounded by the original polygon is the same as the area of the resulting 5 segment by 5 segment square. Since the length of each segment is $56/28 = 2$, the required area is thus $(5 \cdot 2)^2 = 100$.

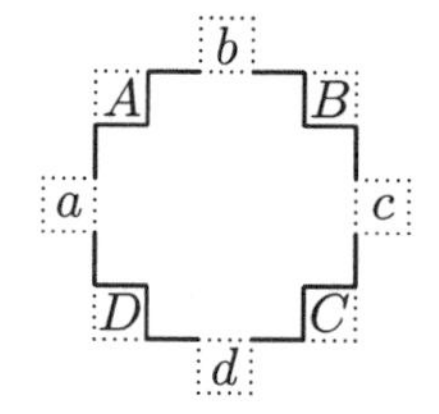

9. **(D)** Since $\angle A = 4\angle B$, and $90° - \angle B = 4(90° - \angle A)$, it follows that

$$90° - \angle B = 4(90° - 4\angle B),$$

or $$15\angle B = 270°.$$

Thus, $\angle B = 18°$.

Note. Observe that the two conditions

(1) *$\angle A$ is k times $\angle B$,*

(2) *the complement of $\angle B$ is k times the complement of $\angle A$,*

imply that

$\angle A$ is the complement of $\angle B$.

That is, if $A \neq B$ then

$$\frac{A}{B} = k = \frac{90° - B}{90° - A} \iff A + B = 90°.$$

Proof:

$$\frac{A}{B} = \frac{90° - B}{90° - A} \iff A(90° - A) = B(90° - B)$$
$$\iff B^2 - A^2 = 90°(B - A)$$
$$\iff B + A = 90°.$$

Thus, we could solve $\angle A + \angle B = 90°$ and $\angle A = 4\angle B$ simultaneously for $\angle B$.

10. **(B)** Since $m(b,c) = b$ and $m(a,e) = a$, we have

$$\begin{aligned} M(M(a,m(b,c)),m(d,m(a,e))) &= M(M(a,b),m(d,a)) \\ &= M(b,a) = b. \end{aligned}$$

11. **(D)** The sum of the surface areas of the three cubes is $6+24+54 = 84$. We minimize the surface area by attaching each cube to the other two along an entire face of the smaller cube. The figure shows how this is possible. Each attachment subtracts twice the area of a face of the smaller cube from the total. The remaining surface area is

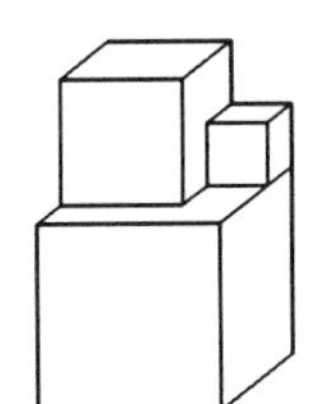

$$84 - 2(1) - 2(1) - 2(4) = 72.$$

12. **(D)**

$$\begin{aligned} \left(i - i^{-1}\right)^{-1} &= \left(i - \frac{1}{i}\right)^{-1} \\ &= \left(\frac{i^2 - 1}{i}\right)^{-1} \\ &= \left(\frac{-2}{i}\right)^{-1} = -\frac{i}{2}. \end{aligned}$$

OR

Since $i^{-1} = -i$, it follows that

$$\left(i - i^{-1}\right)^{-1} = (i - (-i))^{-1} = (2i)^{-1} = 2^{-1}i^{-1} = -\frac{i}{2}.$$

13. **(B)** Let $\angle A = x^\circ$. Then $\angle PCA = x^\circ$ since $AP = PC$. By the exterior angle theorem,

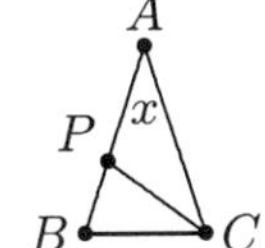

$$\angle BPC = \angle A + \angle PCA = 2x^\circ.$$

Since $PC = CB$, it follows that $\angle B = 2x^\circ$.
Thus $\angle ACB = 2x^\circ$ since $AB = AC$. Summing the angles in $\triangle ABC$ yields $x^\circ + 2x^\circ + 2x^\circ = 180^\circ$, or $\angle A = x^\circ = 36^\circ$.

Comment. Did you notice that since $\angle ACB = 2x^\circ$ and $\angle ACP = x^\circ$, it follows that $\overline{PC}$ bisects $\angle ACP$?

14. **(B)** Since $40 = 20 + 100(1/5)$, there are 101 terms. In an arithmetic series, the sum is the number of terms times the average of the first and last terms. Thus the desired sum is

$$101 \cdot \left(\frac{20+40}{2}\right) = 101 \cdot 30 = 3030.$$

OR

There are 101 terms. Write the sum of the last fifty terms in reverse order under the first fifty:

$$\begin{array}{l} 20 + 20\frac{1}{5} + 20\frac{2}{5} + \cdots + 29\frac{3}{5} + 29\frac{4}{5} + \\ \hfill 30 + \\ 40 + 39\frac{4}{5} + 39\frac{3}{5} + \cdots + 30\frac{2}{5} + 30\frac{1}{5} \\ \hline \underbrace{60 + 60 \quad + 60 \quad + \cdots + 60 \quad + 60}_{50 \text{ terms}} \quad + 30 \end{array}$$

Thus the sum is $50(60) + 30 = 3030$.

Comment. There is a solution similar to the above using decimals instead of fractions.

OR

Since

$$\frac{1}{5} + \frac{2}{5} + \frac{3}{5} + \frac{4}{5} = 2,$$

regroup the given expression as indicated to find the sum:

$$\begin{aligned} (5(20) + 2) + (5(21)+2) + \cdots + (5(39) + 2) + 40 \\ &= 5(20 + 21 + \cdots + 39) + 20(2) + 40 \\ &= 5\left(20 \cdot \frac{20+39}{2}\right) + 80 \\ &= 2950 + 80 = 3030. \end{aligned}$$

OR

The sum is

$$\frac{100}{5} + \frac{101}{5} + \cdots + \frac{200}{5} = \frac{100 + 101 + \cdots + 200}{5}.$$

The sum of the 101 terms in the numerator is

$$101\left(\frac{100+200}{2}\right)=101(150),$$

so the desired sum is $\dfrac{101(150)}{5}=101(30)=3030.$

15. **(B)** The squares of only two integers in $\{1,2,3,\ldots,10\}$ have odd tens digits, $4^2=16$ and $6^2=36$. Since

$$(n+10)^2=n^2+(20n+100)$$

and the tens digit in $20n+100$ must be even, it follows that the tens digit in $(n+10)^2$ will be odd if and only if the tens digit in n^2 is odd. Inductively, we conclude that only numbers in $\{1,2,3,\ldots,100\}$ with units digit 4 or 6 will have squares with an odd tens digit. There are exactly $10\times 2=20$ such numbers.

OR

Since neither 0^2 nor 100^2 have odd tens digits, we replace the given set with $\{0,1,2,\ldots,99\}$, and count the numbers n of the form

$$n=10m+d,\qquad d=0,1,\ldots,9,\qquad m=0,1,\ldots,9$$

for which n^2 has an odd tens digit. Compute

$$n^2=(10m+d)^2=10(10m^2+2md)+d^2.$$

Since $10m^2+2md=2(5m^2+md)$ is even and its units is added to the tens digit of d^2 to form the tens digit of n^2, it follows that the tens digit of n^2 will be odd if and only if the tens digit of d^2 is odd. There are two digits, $d=4$ and $d=6$, for which the tens digit of d^2 is odd. Since there are 10 choices for m to pair with these two choices for d, there are $2\times 10=20$ integers n in the set whose squares have odd tens digits.

16. **(B)** Let r be the number of red marbles and n the total number of marbles originally in the bag. Then

$$\frac{r-1}{n-1}=\frac{1}{7}\qquad\text{and}\qquad\frac{r}{n-2}=\frac{1}{5}.$$

Therefore $7r - 7 = n - 1$ and $5r = n - 2$. Solve

$$7r - n = 6$$

$$5r - n = -2$$

simultaneously to find that $n = 22$.

17. **(D)** Let O be the center of the circle and rectangle, and let the circle and rectangle intersect at A, B, C and D as shown. Since $AO = OB = 2$ and the width of the rectangle $AB = 2\sqrt{2}$, it follows that $\angle AOB = 90°$. Hence $\angle AOD = \angle DOC = \angle COB = 90°$. The sum of the areas of sectors AOB and DOC is $2\left(\frac{1}{4}\left(\pi 2^2\right)\right) = 2\pi$. The sum of the areas of isosceles right triangles AOD and COB is $2\left(\frac{1}{2}\cdot 2^2\right) = 4$. Thus, the area of the region common to both the rectangle and the circle is $2\pi + 4$.

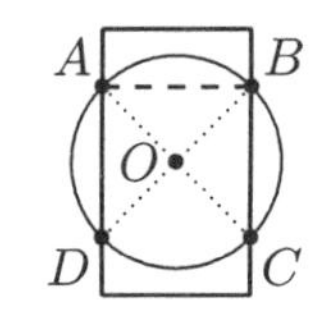

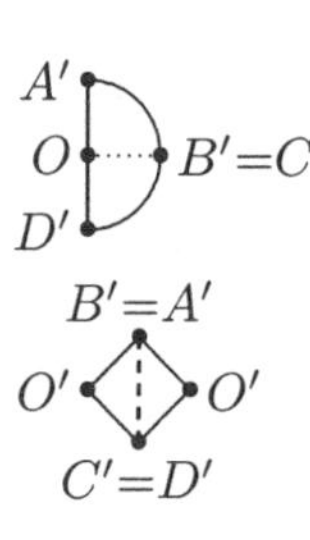

Note. The sketch shows that sectors AOB and COD can be arranged to form a semicircle with area $\frac{1}{2}\pi 2^2 = 2\pi$ and the isosceles right triangles can be arranged to form a square with area $2^2 = 4$.

Comment. This problem can be more difficult if the radius and dimensions of the rectangle are chosen in a more arbitrary manner.

18. **(C)** Since $180° = \angle A + \angle B + \angle C = \angle A + 4\angle A + 4\angle A$, it follows that $\angle A = 20°$. Therefore, arc $BC = 2\angle A = 40°$, which is $1/9$ of $360°$. Thus the polygon has 9 sides.

OR

If $\angle C$ is partitioned into four angles congruent to $\angle A$, the four chords associated with the arcs subtended by these angles will be congruent to $\overline{BC}$. These four chords plus four obtained analogously from $\angle B$, together with $\overline{BC}$, form the $n = 9$ sides of the inscribed regular polygon.

Note. In general, if $\angle B = \angle C = k\angle A$, then $n = 2k + 1$.

19. **(C)** The largest set of disks that does not contain ten with the same label consist of all 45 disks labeled "**1**" through "**9**", and nine of each

of the other 41 types. Hence, the maximum number of disks that can be drawn without having ten with the same label is

$$45 + 41(9) = 414.$$

The 415th draw must result in ten disks with the same label, so 415 is the minimum number of draws that guarantees at least ten disks with the same label.

20. **(B)** Because x, y, z forms a geometric sequence with common ratio r, we have

$$y = xr \quad \text{and} \quad z = yr = xr^2.$$

Because $x, 2y, 3z$ forms an arithmetic sequence, the common difference is

$$3z - 2y = 2y - x.$$

Substitute for y and z in this equation and solve the result, noting that $x \neq y$ implies that $x \neq 0$ and $r \neq 1$:

$$3xr^2 - 2xr = 2xr - x \iff 3r^2 - 4r + 1 = 0,$$
$$\iff (3r - 1)(r - 1) = 0$$
$$r = 1/3.$$

Note that $r = 1$ was impossible since $x \neq y$.

Comment. Whenever x, y, z is a nontrivial geometric sequence and x, ay, bz forms an arithmetic sequence, then the common ratio $r = 1/b$.

21. **(C)** An odd integer the sum of whose digits is 4 must have units digit 1 or 3. This is a list of all the odd positive integers with non-zero digits, the sum of whose digits is 4:

$$13, \ 31, \ 121, \ 211, \ 1111.$$

Since 13, 31 and 211 are primes, $121 = 11^2$ and $1111 = 11 \cdot 101$ are the two counterexamples.

Note. To verify that 211 is prime, we first note that it is odd, its units digit is not 5, $2+1+1$ is not divisible by 3, $2-1+1$ is not divisible by 11, and $211 < 17^2$. Therefore we need to divide 211 by only 7 and 13 to see that it is prime.

22. **(C)** The two end chairs must be occupied by students, so the professors have seven middle chairs from which to choose, with no two adjacent. If these chairs are numbered from 2 to 8, the three chairs can be:

$$(2,4,6),\ (2,4,7),\ (2,4,8),\ (2,5,7),\ (2,5,8)$$

$$(2,6,8),\ (3,5,7),\ (3,5,8),\ (3,6,8),\ (4,6,8).$$

Within each triple, the professors can arrange themselves in 3! ways, so the total number is $10 \times 6 = 60$.

OR

Imagine the six students standing in a row before they are seated. There are 5 spaces between them, each of which may be occupied by at most one of the 3 professors. Therefore, there are $P(5,3) = 5 \times 4 \times 3 = 60$ ways the three professors can select their places.

23. **(E)** The area of the region is $3^2 + (2)(1) = 11$. Label the vertices as indicated in the figure. Since the area of trapezoid $OABC$ is $(2+5)/2 < 11/2$ and the area of triangle ODE is $3^2/2 < 11/2$, it follows that the desired line, $y = mx$, intersects the line $x = 3$ at some point $(3, 3m)$, where $1 < 3m < 3$. The area of the trapezoid above the line $y = mx$ is

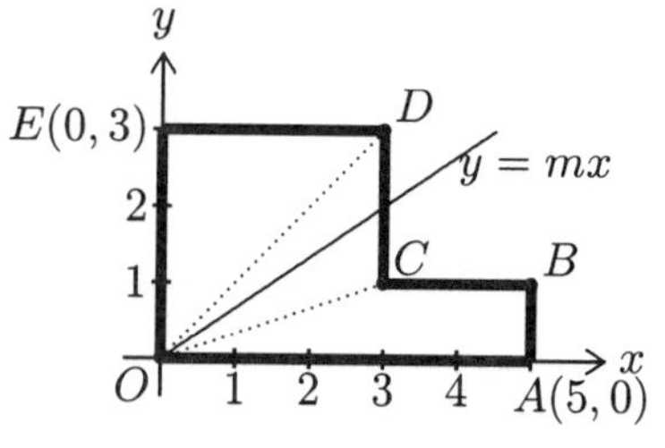

$$\frac{[3 + (3 - 3m)]}{2}(3)$$
$$= \frac{18 - 9m}{2} = \frac{11}{2}.$$

Solve this equation to find $m = 7/9$.

OR

The area to the right of the line $x = 3$ in the L-shaped region is 2. Since $\frac{2}{3} \times 3 = 2$, the area above the line $y = 7/3$ is also 2. The diagonal of the rectangle which remains when these two rectangles of area 2 are discarded is the line which bisects the area of the L-shaped region. This diagonal connects the origin with $(3, 7/3)$ and has slope $7/9$.

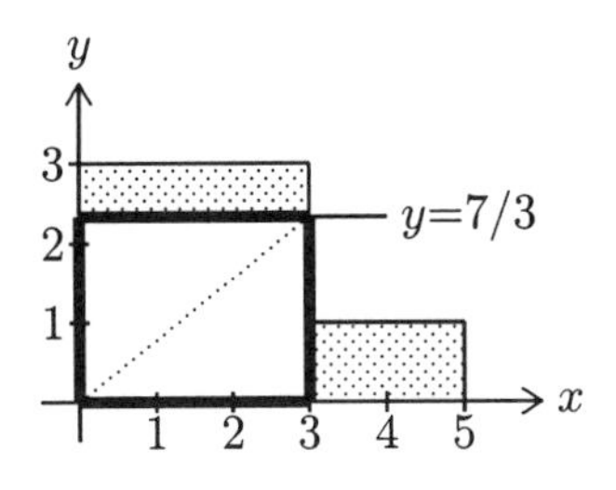

OR

Label the vertices as shown, and note that $\overline{OD}$ partitions square $OFDE$ into two triangles of equal area. We seek G on $\overline{CD}$ so $[ODG] = [FABC]/2 = 1$. Also

$$1 = [ODG] = \frac{1}{2}(DG)(OF) = \frac{(DG)(3)}{2},$$

so $DG = 2/3$ and

$$FG = 3 - \frac{2}{3} = \frac{7}{3}.$$

Thus line $\overline{OG}$ has slope $(7/3)/3 = 7/9$.

Comment. These three solutions certainly illustrate the nonuniqueness for the location of auxiliary lines to solve a geometry problem.

24. **(C)** Since the mean of the 5 observations is 10, their sum must be 50. Since the median is 12, one observation to be 12, two others can be no more than 12, and the remaining two must be at least 12. If a maximal observation is increased by x, the sum of those no larger than 12 must be reduced by x in order to keep the mean at 10. However, this expands the range. Thus the minimum range will occur when three observations are 12 and the remaining two observations are equal and sum to $50 - 3(12) = 14$. Hence the sample $7, 7, 12, 12, 12$ minimizes the range, and the smallest value that the range can assume is $12 - 7 = 5$.

25. **(A)** If $x > 0$, then $x + y = 3$ and $y + x^2 = 0$. Eliminate y from these simultaneous equations to obtain $x^2 - x + 3 = 0$, which has no real roots. Equivalently, note that the graphs of $y = 3 - x$ and $y = -x^2$ do not intersect.

If $x < 0$, then we have $-x + y = 3$ and $-y + x^2 = 0$, which have a simultaneous real solution, so $x - y = -(-x + y) = -3$.

Note. One need not solve $x^2 - x - 3 = 0$ and discover that

$$(x, y) = \left(\frac{1 - \sqrt{13}}{2}, \frac{7 - \sqrt{13}}{2}\right),$$

to find the answer to this problem.

OR

Sketch

$$y = 3 - |x|$$

$$\text{and } y = \frac{-x^3}{|x|} = \begin{cases} x^2 & \text{if } x < 0 \\ -x^2 & \text{if } x > 0. \end{cases}$$

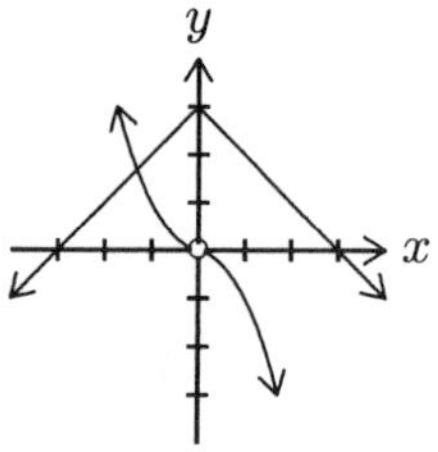

Note that the graphs cross only on the half-line $y = x + 3$, $x < 0$. Therefore $x - y = -3$.

26. **(A)** The measure of each interior angle of a regular k-gon is $180° - (360°/k)$. In this problem, each vertex of the m-gon is surrounded by one angle of the m-gon and two angles of the n-gons. Therefore,

$$\left(180° - \frac{360°}{m}\right) + 2\left(180° - \frac{360°}{n}\right) = 360°,$$

$$180°\left[\left(1 - \frac{2}{m}\right) + 2\left(1 - \frac{2}{n}\right)\right] = 180°[2],$$

$$\left(1 - \frac{2}{10}\right) + 2\left(1 - \frac{2}{n}\right) = 2.$$

Solve this equation to find that $n = 5$.

Note. The equation

$$\left(1 - \frac{2}{m}\right) + 2\left(1 - \frac{2}{n}\right) = 2$$

may be written as

$$(mn - 2n) + (2mn - 4m) = 2mn$$

$$(m-2)(n-4) = 8 = \begin{cases} 1 \cdot 8 \\ 2 \cdot 4 \\ 4 \cdot 2 \\ 8 \cdot 1 \end{cases}$$

which shows its only solutions in positive integers to be $(m, n) = (3, 12)$, $(4, 8)$, $(6, 6)$, and $(10, 5)$.

OR

Each interior angle of a regular decagon measures $(180° - 36°)$. The interior angles of the two n-gons at one of its vertices must fill $360° - (180° - 36°) = 216°$. The regular polygon each of whose interior angles measures $216°/2 = 108°$ is the pentagon, so $n = 5$.

27. **(D)** Let the total number of kernels be $3n$, so that there are $2n$ white kernels and n yellow kernels. Then

$$\frac{1}{2}(2n) + \frac{2}{3}(n) = n + \frac{2n}{3} = \frac{5n}{3}$$

of all $3n$ kernels will pop. Since n white kernels will pop, the probability that the popped kernel was white is

$$\frac{n}{5n/3} = \frac{3}{5}.$$

OR

Make a probability tree diagram:

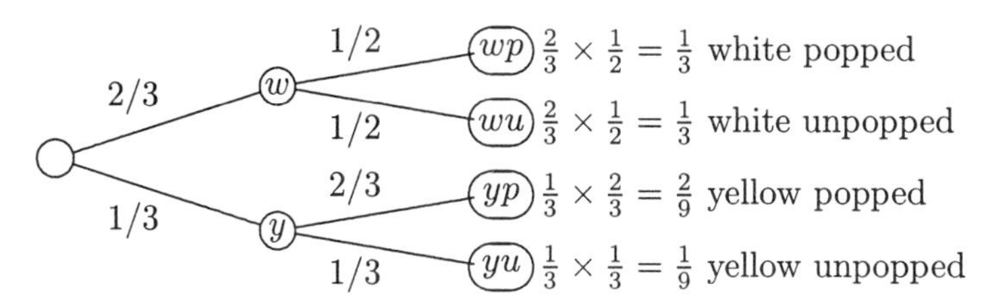

Since

$$\frac{1}{3} + \frac{2}{9} = \frac{5}{9}$$

of the kernels popped and $1/3$ of the kernels are white and popped, the probability that the popped kernel was white is

$$\frac{1/3}{5/9} = \frac{3}{5}.$$

OR

Make a diagram as indicated, letting areas represent the probabilities. The ratio of the area shaded with line segments to the area shaded with segments or dots is

$$\frac{\left(\frac{1}{2}\right)\left(\frac{2}{3}\right)}{\left(\frac{1}{2}\right)\left(\frac{2}{3}\right)+\left(\frac{2}{3}\right)\left(\frac{1}{3}\right)} = \frac{3}{5}.$$

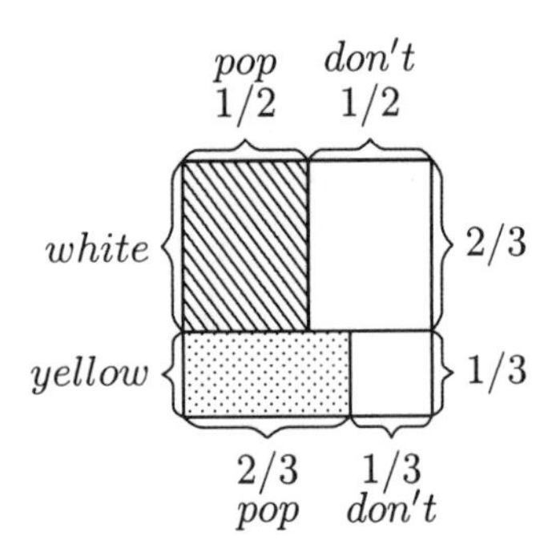

OR

Use a Venn diagram where the universal set is the set of kernels, $\mathbf{W}$ is the set of white kernels, and $\mathbf{P}$ is the set of kernels which will pop. Let x be the number of kernels in $\mathbf{W} \cap \mathbf{P}$ and $\mathbf{W} - \mathbf{P}$, and let y be the number of kernels not in $\mathbf{W} \cup \mathbf{P}$. Then there are $2y$ kernels in $\mathbf{P} - \mathbf{W}$. Thus, we are given

$$\frac{x+x}{x+x+2y+y} = \frac{2}{3},$$

so $x = 3y$. The probability that a kernel in $\mathbf{P}$ is also in $\mathbf{W}$ is

$$\frac{x}{x+2y} = \frac{3y}{3y+2y} = \frac{3}{5}.$$

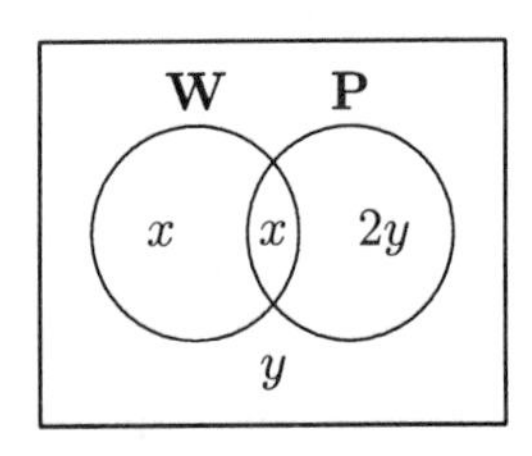

Note. This is an application of Bayes' Theorem.

28. **(C)** Write the equation of the line in the *two-intercept form*:

$$\frac{x}{p} + \frac{y}{b} = 1, \quad \text{where } p \text{ is prime and integer } b > 0.$$

Substitute $x = 4$ and $y = 3$ to obtain

$$\frac{4}{p} + \frac{3}{b} = 1 \qquad \text{or} \qquad b = \frac{3p}{p-4} = 3 + \frac{12}{p-4}.$$

Since b is a positive integer, $(p-4)$ must be a positive divisor of 12. There are only two such primes, $p = 5$ and $p = 7$. Therefore, there are two lines with the requested properties,

$$\frac{x}{5} + \frac{y}{15} = 1 \quad \text{and} \quad \frac{x}{7} + \frac{y}{7} = 1.$$

OR

Let p and b be the x- and y-intercepts of such a line. Since this line through $(4,3)$ intersects the positive y-axis, its x-intercept, $p > 4$. Analogously, $b > 3$. Since the points $(p,0)$, $(4,3)$ and $(0,b)$ are collinear, by computing the slope of the line in two different ways, we find

$$\frac{b-3}{0-4} = \frac{0-3}{p-4},$$
$$(p-4)(b-3) = 12.$$

Thus, $(p-4)$ must be one of the divisors $d = 1, 2, 3, 4, 6$ or 12, of 12; and $p = 4 + d$ must be an odd prime. Testing the only two odd divisors d of 12, we find that there are only two such primes, $p = 4+1 = 5$ and $p = 4+3 = 7$.

OR

Since both intercepts must be positive, the lines

$$\frac{x}{p} + \frac{y}{b} = 1$$

with the desired properties must have negative slope. Thus, the integer b is larger than 3, so $b \geq 4$. Similarly, $p \geq 5$. Draw the line from $(0,4)$ through $(4,3)$ to see that $p \leq 16$.

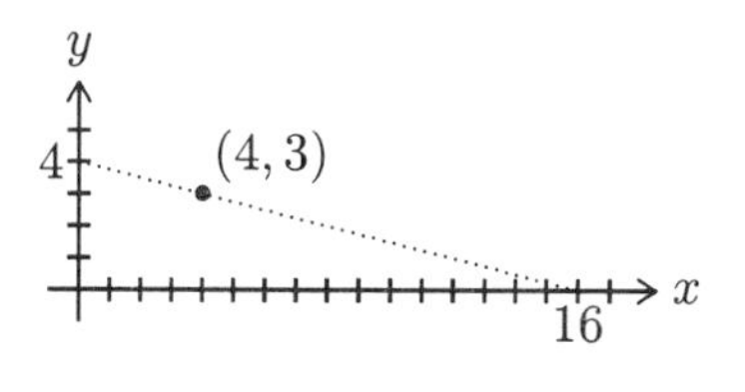

There are four primes between 4 and 16: 5, 7, 11 and 13.

If $p = 5$, the slope is -3 and $b = 15$.

If $p = 7$, then the slope is -1 and $b = 7$.

If $p = 11$, then the slope is $-3/7$ and the y-intercept is $33/7$, which is not a suitable integer value for b.

Similarly, $p = 13$ yields a slope of $-1/3$ and y-intercept $13/3$, which is not an integer.

Thus, there are two such lines.

29. **(A)** Draw and label the figure as shown, where O is the center of the circle. By the definition of radian, $\angle BOC = 1$, so $\angle BAC = 1/2$ and $\angle BAE = \frac{1}{2}\angle BAC = 1/4$. Since $\overline{BE} \perp \overline{EA}$, it follows that

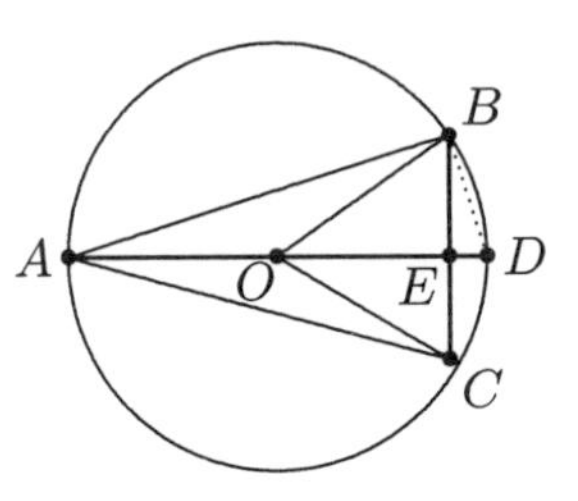

$$\frac{AB}{BC} = \frac{AB}{2BE} = \frac{1}{2}\frac{AB}{BE}$$
$$= \frac{1}{2}\csc\angle BAE = \frac{1}{2}\csc\frac{1}{4}.$$

OR

Since $\angle BAC = 1/2$ and the interior angles in isosceles $\triangle ABC$ have sum π, it follows that

$$\angle ACB = \frac{1}{2}\left(\pi - \frac{1}{2}\right) = \frac{\pi}{2} - \frac{1}{4}.$$

The length of a chord subtended by an inscribed angle β in a circle of radius r is $2r\sin\beta$, (*Why?*†) so

$$AB = 2r\sin\angle ACB = 2r\sin\left(\frac{\pi}{2} - \frac{1}{4}\right) = 2r\cos\frac{1}{4}.$$

$$BC = 2r\sin\angle BAC = 2r\sin\frac{1}{2} = 4r\sin\frac{1}{4}\cos\frac{1}{4},$$

Therefore, $$\frac{AB}{BC} = \frac{2r\cos\frac{1}{4}}{4r\sin\frac{1}{4}\cos\frac{1}{4}} = \frac{1}{2\sin\frac{1}{4}} = \frac{1}{2}\csc\frac{1}{4}.$$

OR

By the *Law of Sines*,

$$\frac{AB}{BC} = \frac{\sin C}{\sin A}.$$

But $2\angle C + \angle A = \pi$, so

$$\angle C = \frac{\pi}{2} - \frac{\angle A}{2} \quad \text{and} \quad \sin C = \cos\frac{A}{2}.$$

---

† The length of the chord subtended is the same for all orientations of the inscribed angle β because the length of the arc subtended does not change. Therefore orient $\angle\beta$ so one side corresponds to the diameter of length $2r$. Since angles inscribed in semicircles are right angles, the chord-length is $2r\sin\beta$. For example, note in the diagram for the first solution that $BD = AD\sin\angle BAD$.

Since $\angle A = 1/2$,

$$\frac{AB}{BC} = \frac{\sin C}{\sin A} = \frac{\cos \frac{A}{2}}{2\sin \frac{A}{2} \cos \frac{A}{2}} = \frac{1}{2\sin \frac{A}{2}}$$
$$= \frac{1}{2}\csc \frac{A}{2} = \frac{1}{2}\csc \frac{1}{4}.$$

30. **(C)** When n dice are rolled, the sum can be any integer from n to $6n$. The sum $n+k$ can be obtained in the same number of ways as the sum $6n-k$, and this number of ways increases as k increases from 0 to $\lfloor 5n/2 \rfloor$. Minimize $S = n+k$ by choosing n and k as small as possible with $6n - k = 1994$. Since the least multiple of 6 that is greater than or equal to 1994 is $1998 = 6(333)$, S is smallest when $n = 333$ and $k = 4$. Consequently, $S = n + k = 337$.

OR

On a standard die, 6 and 1, 5 and 2, and 4 and 3 are on opposite sides. To obtain a sum of 1994 with the most sixes on the top faces of the dice requires that 332 sixes and 1 two face up. Then 332 ones and 1 five will face down, and $332 + 5 = 337$.

OR

Each roll of the n dice can be represented by the n-tuple $(a_1, a_2, \ldots, a_n)$. There is a one-to-one correspondence between the rolls $(a_1, a_2, \ldots, a_n)$ and the rolls $(7-a_1, 7-a_2, \ldots, 7-a_n)$. Thus, for all x, the probability of obtaining a sum of x equals the probability of obtaining a sum of $7n - x$.

In this case, we have $7n - S = 1994 = 7{\cdot}285 - 1$, so $S - 1 = 7(n - 285)$. Thus, $S - 1$ is a multiple of 7; that is, for some positive integer k, $S = 7k + 1$. In order to be able to obtain a sum of 1994,

$$n \geq \frac{1994}{6} = 332\frac{1}{3}, \quad \text{so} \quad n \geq 333,$$

and since $S \geq n$, $S \geq 333 = 7(47) + 4$. Since S must be of the form $7k + 1$, the minimum value of S is $7 \cdot 48 + 1 = 337$.

OR

When n dice are rolled, the sum can be any integer from n to $6n$. For any j between n and $6n$, the number of ways the sum j can appear is the coefficient of x^j in the generating function

$$P(x) = (x + x^2 + x^3 + x^4 + x^5 + x^6)^n. \qquad (Why?)$$

By symmetry, in $P(x)$ the coefficient of x^j equals the coefficient of x^{7n-j}. We leave it as an exercise to show that for $n > 1$, the coefficients in $P(x)$ are strictly increasing for $j = n, n+1, \ldots, \lfloor 7n/2 \rfloor$. Therefore, there is only one value of $S \neq 1994$ for which the coefficients of x^S and x^{1994} are equal. In this case we have $7n - S = 1994$, or $S = 7n - 1994 = 7(n - 285) + 1$. Hence the minimum value of S corresponds to the smallest allowable value of n. Since $1994 \leq 6n$, it follows that $n \geq 333$. Therefore, $S = 7(333 - 285) + 1 = 337$.

An Insider's Look at the AHSME Problems

Where Do Ideas for Problems Come From?

Some problems undoubtedly come from mathematical research. The drawing originally submitted with the solution to problem 22 on the 43rd AHSME was

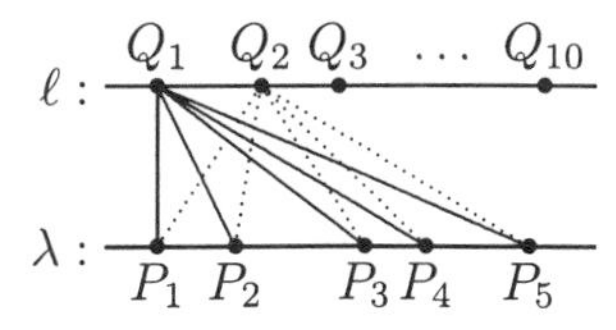

which suggests the idea for this problem might have been generated as someone stared at the typical picture for the complete bipartite graph known as $\mathbf{K}_{10,5}$. The committee made the line containing the P_i perpendicular to the line containing the Q_i because it seemed easier to describe the problem in that orientation without a sketch.

Driving might start the creative juices to flow in some problem posers. Problem 12 on the 40th AHSME, which asks for an estimate of the number of cars on a section of highway by counting the cars we pass for five minutes, was suggested by someone who lives close to a straight interstate highway in the great plains. The idea for problem 5 on the 42nd AHSME, in which we find the area of a large arrow, was spawned as the poser was driving home from the office late at night and passed one of those large flashing yellow arrows at a highway construction site.

Problem 9 on the 40th AHSME concerns how many initials in alphabetical order a person could have if the last name began with Z. This problem was proposed by a panel member whose last name begins with a letter near the end of the alphabet about a year after marriage, so it is not difficult to guess what led to this question.

The poser of problem 9 on the 43rd AHSME created the figure so the problem would use the formula for the area of a rhombus. Reviewers worked on the problem and submitted many alternative approaches using formulas for equilateral triangles and trapezoids. Consequently, the original solution using the formula for a rhombus is listed as just one of four approaches to the problem, and is not even the primary solution.

A concept that has been around mathematical recreations for over a century are "repunits", numbers whose base ten expansion consists entirely of 1s. Problem 7 on the 44th AHSME requires the student to find the pattern in the quotient when one particular repunit divides evenly into another.

Very occasionally excess information can be used to make a trivial question into a question usable on a contest. In problem 7 on the 40th AHSME there is a right triangle AHC with M at the midpoint of the hypotenuse $\overline{AC}$. Since the midpoint of the hypotenuse is the center of the circumcircle, $\triangle AHM$ and $\triangle MHC$ are both isosceles, from which the answer follows immediately. The statement of this problem inserts an extra point B, and tells us only indirectly that M is the midpoint of $\overline{AC}$ by stating that $\overline{BM}$ is a median of $\triangle ABC$. Further extraneous information in this problem are the measure of angles A and B.

Solutions Affect Interest in Problem

Many times the solution that the panel and committee reviews with the problem affects the decision to use the problem or not. Problem 27 on the 40th AHSME, which asks about the number of positive integer solutions to

$$2x + 2y + z = n,$$

was originally submitted and reviewed by the committee with the solution that appears as the alternative solution. Even in that form, the AHSME committee liked the problem. Enthusiasm for this problem might not have been enough for it to make the cut for the 30 problems to be used on this contest had what is now the primary solution, based on the number of lattice points in an isosceles right triangle, not been submitted by one of our reviewers during the review of one of our drafts. The present alternative solution, the original solution we considered, shows how the problem is related to a type of problem commonly discussed in a course on discrete mathematics:

"Show that the number of positive integer solutions of

$$x_1 + x_2 + x_3 + \cdots + x_k = t.$$

is $\binom{t-1}{k-1}$."

See, for example, Example 5 on page 205 in *Applied Combinatorics*, Third Edition, by Alan Tucker, published by John Wiley & Sons, or Exercise 1.4, 7b on page 34 in *Discrete and Combinatorial Mathematics*, Fourth Edition, by Ralph Grimaldi, published by Addison-Wesley. This method is sometimes called the "Stars and Bars" technique.

The presence of the 2s in

$$2x + 2y + z = n$$

distinguishes this AHSME problem from the routine stars and bars exercise. However, it is not far removed, since our alternative solution shows how the AHSME problem reduces to

$$x + y + j = \left\lceil \frac{n}{2} \right\rceil$$

which has, by the above formula,

$$\binom{\lceil \frac{n}{2} \rceil - 1}{3 - 1} = 28$$

solutions.

Problem Difficulty

Not everyone means the same thing by "difficult problem." For some, a difficult problem is one which fewest students answered correctly; for others, one which the most students answered incorrectly; and for still others, one which few students even attempted. Personally, I like to look at the score attained on the problem averaged over a large sample of participants. It is not uncommon for a given AHSME to have a different "most difficult problem" by each of these four criteria.

The committee occasionally attempts to fine tune the difficulty of an AHSME. This can be done in many ways. Geometry problems are more difficult if the figure is not given with the problem. The removal of a couple really good "Gotchas" from the distractors increases the net score on any problem. Different aspects of a problem can be asked. For example, problem 30 on the 45th AHSME asks for the smallest score with the same probability as a score of 1994 if the latter score is possible. A different version of that problem, which would have made that examination easier is:

> If a sum of 1994 can be rolled with n standard 6-sided dice, then the smallest possible value of n is
>
> **(A)** 331 **(B)** 332 **(C)** 333 **(D)** 334 **(E)** 335

In this case the committee went with the version of the problem thought to be more difficult. Since the 45th AHSME set a record with its large number of Honor Roll students, the decision was prudent!

Problems on Applications

Good problems about applications of mathematics are rare because most require more explanation than can reasonably be given on a timed test. Problem 8 on the 42nd AHSME comes close to applied mathematics since its connection to oil spills is obvious.

On the other hand, several problems relate to questions that might concern us in everyday life. Problem 9 on the 40th AHSME asks about how many initials in alphabetical order the Zeta couple can give their newborn. The solution to problem 26 on the 41st AHSME reveals that problem to be a very simple discrete example of inverting averages, which is an analog to the continuous three-dimensional case used in CAT scans.

When deciding whether to increase the number of words to explain more about the application in the statement of the problem, the CAMC invariably prefers the shorter version of the problem because of the timed-test format.

Sometimes a problem derived from an application is easier to word than the actual application that originally suggested the problem. Problem 3 on the 40th AHSME was originally about folding a piece of paper into three parts, something we all do with letters we write. Since a key fact in the problem is the perimeter of each of the three parts, the committee thought it would be clearer to have the paper cut rather than folded.

Convenient numbers have also been preferred over numbers that are more realistic. The numbers in problem 11 on the 42nd AHSME work out nicely, but we have Jill running at a rate close to a four minute mile for the downhill five of her ten kilometers. In editing the problem, when the committee became aware of the speed involved, it changed the original word "jogging" to the word "running," which was used for the test.

Auxiliary Lines

Because examples in geometry texts include only one solution, teachers are asked for a rule for drawing the auxiliary lines to obtain the solution. The multiple solutions to some of the geometry problems in this book show that not infrequently, many different auxiliary lines can be used, so there can be no fixed rule for placing auxiliary lines. The following AHSME geometry problems, listed as (AHSME Number, Problem Number), have two or more distinct diagrams with their solution: $(40,7)$, $(40,21)$, $(42,19)$, $(42,22)$, $(42,23)$, $(43,9)$, $(43,20)$, $(43,25)$, $(44,14)$, $(44,17)$, $(44,27)$, $(45,8)$. Many of the diagrams are associated with more than one approach in the solution.

Fashion in Problems

Once upon a time, when drafting was a part of mathematics and when three-dimensional geometry was emphasized, the following might have been an appropriate AHSME problem. This problem was proposed for one of the AHSMEs and was never voted beyond the initial packet of problems, which is a sign of the times, fashion-wise, in mathematics.

0. The outline of the shape of a monument when viewed from directly above is

   □

   Which of the following could be outlines of the shape of the monument when viewed from the West and North?

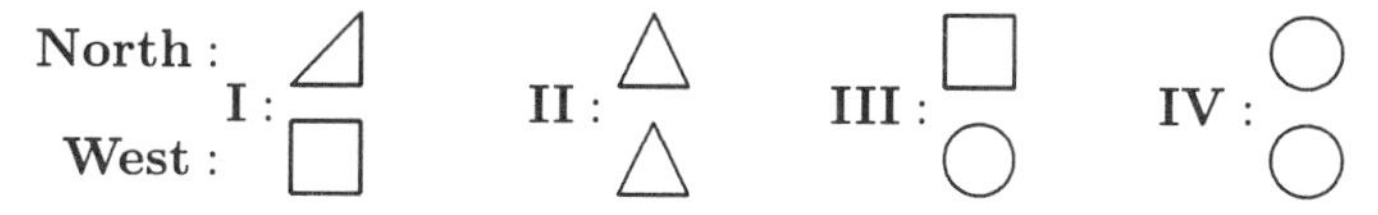

(A) **I** only **(B)** **I** or **II** only **(C)** **I** or **III** only
(D) **I**, **II** or **III** only **(E)** Any of **I**, **II**, **III** or **IV**

The following is its solution:

0. **(E)** The views in **I** could result if the monument were half a cube obtained by passing a plane through opposite edges, those in **II** could result if it were a pyramid with a square base, those in **III** if it were a

cylinder lying on its side, and those in **IV** if it were shaped like the part in both of two cylinders of identical diameter with axes intersecting at right angles.

Problem Clarity

When you know there will be between a third and a half of a million students attempting your questions, you try to write clearly, primarily to make the examination fair, but also to thwart those who might be looking for excuses for their wrong answers. No matter how much care the AHSME committee exercises in writing the test, the results exhibit how ambiguous the English language is, even in mathematics.

Many times a student's interpretation leads to an answer not included among the five choices, and the AHSME committee feels that the rules, which state that one and only one of the choices is correct, should have alerted the student to think of an alternative interpretation. For example, the distractor $\binom{26}{3} = 2600$ was deliberately not used for problem 9 on the 40th AHSME to avoid the (legally valid) possibility that Mr. and Mrs. Zeta did not want their child to share their last name.

Problem 6 on the 41st AHSME asks for the number of lines in the plane that are 2 units from A and 3 units from B. The poser of the problem included with the statement of the problem the fact that distance from a point to a line is measured along the segment through the point perpendicular to the line. When the committee discussed this problem, they decided that the definition of the distance from a point to a line was well known, and therefore not necessary to include in the statement of the problem. We were wrong! Our correspondence indicates that lots of students thought there were an infinite number of lines 2 units from A and 3 units from B.

Problem 5 of the 40th AHSME shows 30 toothpicks in the accompanying diagram used to depict the 20 by 10 grid discussed in the problem. In this case, one of the answer choices was 30, and you know there were those who argued that should have been the correct answer.

Most Problems are Fresh

While the committee strives for freshness in all its problems, it realizes that it may have to compromise for the first half dozen or so problems on the AHSME in order to include some problems accessible by all participants. On the other hand, since some high-scoring students are very well read, every reasonable effort is made to keep the more difficult problems independent from any published source. An exception to this occurred with problem 15 on the 44th AHSME. While this test was being printed, one of our panel members discovered the identical problem in *Mathematics of Choice, How to Count without Counting* by Ivan Niven which is #15 in the MAA New Mathematical Library Series. This book was originally published in 1965 by Random House. The problem in question is exercise 14 in Problem Set 1 on page 6 of the book. When the duplication was found, the committee decided not to stop the presses and change the AHSME because such changes could have delayed shipment of the tests in addition to incurring unexpected costs. An important factor in the decision was the number of years since the reference was published, giving more students equal access to it. An occasional lapse like this is probably not all bad if it encourages students to search the literature to learn more mathematics lest there be another such lapse. The American Mathematics Competitions pride themselves on their fresh problems and try to keep such duplication of already published problems at a minimum. Would publishers be more happy with us because they might sell more books if we did not adhere so strictly to this principle?

Sensitivity and Problems

The committee tries to be sensitive in the phrasing of its problems. "The price of a dress is half off . . ." is avoided because students who do not read the first three words of the sentence might snicker during the test. Because we could never purge all the pronouns from an applications problem about a boss and secretary,† the problem never was used. Problem 5 on the 44th AHSME began its life with us as a problem about percent increases on a baseball and glove, but the committee changed the equipment to cycling, a sport enjoyed more uniformly by both genders. When problem 16 on the 41st AHSME was proposed, it simply began "At a party" and then described the differences in handshaking responsibilities based on gender. The customs mentioned made a good mathematics problem, but

† "The boss he . . . and the secretary she . . ." would never do! Yet, because we feared that too many participants might be steeped in stereotypes, we did not use "The boss she . . . and the secretary he . . ." lest these participants be distracted from their mathematical thoughts while writing the examination.

were certainly not in vogue in today's world, if they were valid anytime. The committee decided to save the problem by using "At one of George Washington's parties" as the opening phrase since everyone knows that the gender-blindness of social customs is only a recent phenomenon. Problem 22 on the 45th AHSME required the use of three names of teachers. After tiring of the arguments of proponents of two female names and one male name versus the opposite distribution of genders, the committee decided to situate the problem in a college setting since the "Professor" title is gender-independent.

When problem 21 on the 40th AHSME was proposed, the committee knew that the flag described bore some similarity to the flag of the Confederacy during the Civil War. No one checked whether the color scheme on that flag was the same as the one described in this problem. To the committee, it was just a good geometry problem that might have some relationship to some real flag. Apparently this problem gave offense in some quarters, and if the presence of this problem on the examination seemed inappropriate to anyone, the committee offers its sincere apologies. As a result of the appearance of this problem on the AHSME, each year two of the reviewers of the penultimate draft review it solely for issues of sensitivity.

Walter Mientka has been associated with the American Mathematics Competitions as its Executive Director since 1976. His name is printed on every publication associated with our examinations, including the test booklets themselves, so everyone knows his association with our work. Generations of children have perused "Where's Waldo?" books. The idea to use Walter's name in one problem on each AHSME began in drafts of problem 5 on the 45th AHSME. As we reviewed our drafts, Walter was the one confusing division with multiplication and subtraction with addition. However, the sensitivity reviewers insisted on changing to a gender-neutral name, Pat. The idea to introduce "Where's Walter?" problems as an artificial link from one AHSME to the next was delayed as a result of that decision.

AHSME Constants

Any institution of long standing has traditions that must be honored. Some traditionalists become vociferous when those traditions are questioned. Since the AHSME is approaching its fiftieth year, it comes with a fair number of traditions that we dare not question:

- No problem may appear on any of the American Mathematics Competitions if it can be solved more easily using calculus than without.
- The examination must be multiple-choice with exactly five choices.
- Unless a problem asks "Which is the largest [smallest] of the following," the answer choices must be listed in increasing order if they are numbers, or alphabetical order if they are words.
- Geometry problems must be diagram independent. Even when a diagram is included with a problem, all usable assumptions must be stated in words in the problem.

These first two criteria were practical in 1949 when the AHSME began because of the mathematics curriculum in those days and because of the large number of tests that had to be hand-scored. If you look closely, you might find an occasional question in which the committee overlooked the last two criteria.

Since the study of calculus begins with the limit concept, limit problems have never been formally used on any of the American Mathematics Competitions. Problem 25 on the 42nd AHSME would probably make more sense asking

$$\lim_{n\to\infty} P_n = ?$$

but to avoid limits formally, we asked for an approximation to P_{1991}.

When the AHSME began, students learned calculus in the second year of college. Currently hundreds of thousands of students are exposed to some calculus in high school. The role of the AHSME and the other American Mathematics Competitions now seems to be to assure that the rich traditional topics of high school mathematics are not dropped in the rush to get students to calculus.

Multiple Choice

Multiple choice tests are currently much maligned. Such tests are efficient. Lots of the criticism toward teacher-generated multiple-choice tests is well-deserved. For the AHSME, much criticism is thwarted by well-designed tests which go through many drafts with dozens of reviewers suggesting new distractors and criticizing others.

Multiple choice tests ease the burden caused by "simplify" not being well-defined. For arithmetic terms, "simplify" generally means to write the expression in a way that is easy to evaluate. Because long division by hand using a many-digit denominator is computationally intensive, before today's calculators were invented square roots in the denominator were to be avoided, so $10\sqrt{3}/3$ would be a better form for this number than $10/\sqrt{3}$. Today we might define "simple" in terms of the number of key-strokes needed to obtain an approximation on the calculator, so an argument can be made that all the answer choices to problem 25 on the 43rd AHSME, including $8/\sqrt{3}$ and $10/\sqrt{3}$, are written in simplified form.

The AHSME is translated into many languages and used around the world. A couple weeks before the 41st AHSME was scheduled to be given, an email message came to the chair from the person in charge of the translation to be used in Israel. It was thought there was an error in the wording of problem 4 because "is closest to" was used in the statement of the problem and the actual answer 2 was one of the choices. Since any number is closer to itself than any other number, there was no mathematical difficulty with this problem. In fact, wording suggesting that the exact answer might not be one of the choices listed was used deliberately on several questions in the six AHSMEs covered in this book in the hopes that suggestion would help prevent some of the guessing that is inherent on any multiple choice test.

Questions with "None of these" as the fifth distractor were common on the AHSMEs before the six covered in this book. The committee writing these six contests felt that questions using this distractor were never as valid, so the frequency of such questions decreased dramatically. To use such a distractor validly, it should be the correct choice an average of one time in five. In those times when it is the correct choice, many students will get that question correct for the wrong reason.

AHSME Variables

It might be surprising to consider some aspects of the AHSME that have changed through the almost 50-year history of the AHSME. These include:

- Its name.
- The number of problems on the AHSME.
- The formula for determining the student's AHSME score.
- The date of the AHSME.
- The use of calculators on the AHSME.

You can easily detect where these changes occurred with a glance through Contest Problem Books I through V. The only significant change between the 40th and 45th AHSMEs, covered in this book, was the rule on the use of calculators.

Inside Jokes

A committee member brought a campus newspaper to one of our meetings to show us a blatant first-page mathematical error. The paper decried the fact that the college was raising its costs 15% because they had announced that tuition would rise 5% and fees would rise 10%. This led to problem 5 on the 44th AHSME in which the prices of a bicycle and helmet rise 5% and 10%, respectively, and one of the distractors is 15%. The data shows that 15% was chosen by 0.09% of the AHSME Honor Roll students, so these select high school students exhibited better mathematical knowledge than the college newspaper editors.

To help explain the solution to problem 30 on the 40th AHSME we needed two names, one male and one female. The new committee chair that year was from John Carroll University, so the names chosen for the solution were John and Carol.

Leo J. Schneider
John Carroll University
AHSME Chair, 1988–1994

Mathematical Problem Solving for Competitions

The study of mathematics is the same, whether it be for a mathematics class, for one's pure enjoyment in gaining new insights, or for competitions. The use of certain mathematical tools may be encountered more frequently in mathematics competition problems than in text book problems. However, responsible organizers of mathematics competitions try to align the topics they emphasize reasonably well with current curricula. For example, statistics was becoming a more important topic in the secondary schools during the years of the competitions in this book. Note the development from problems 24 on the 41st AHSME and 16 on the 42nd AHSME which are about weighted averages, problems that are actually more algebraic than statistical, to problem 24 on the 45th AHSME which presumes knowledge of fundamental statistical terms.

Tools for Mathematics Competitions

The tools for doing well on the AHSME and AIME are all the topics in a pre-calculus curriculum including some elementary probability, statistics, discrete mathematics, and number theory. The creative aspect of problem solving on competitions lies in the skill of knowing which of these tools to select for which problems. Facility with the use of these tools and their selection can be gained by entering mathematics competitions and by practicing on contests such as those in this book.

Good competitors on timed tests practice to the point where they recognize the use of certain sequences of these tools. They can write down the results without any intermediate writing, almost as if they have the result memorized. One very simple example might be the formula for the **area of an equilateral triangle** of side s. Those experienced in mathematics competitions can picture the 30°-60°-90° triangle that yields the length of the altitude $s\sqrt{3}/2$, mentally substitute this into the formula for the area of a triangle, and obtain the formula for the area, $s^2\sqrt{3}/4$.

Many textbook problems must be written to test the proper acquisition of one new concept. Many competition problems are more difficult simply because their solution requires the use of tools from diverse areas. The first of the following examples is problem 19 from the 45th AHSME, and the second was proposed for the AHSME but never used.

Example 1. From the list

$$1, \underbrace{2, 2}_{2}, \underbrace{3, 3, 3}_{3}, \ldots, \underbrace{50, 50, 50, \ldots, 50}_{50}$$

what is the minimum we must select to guarantee that the selection contains at least ten of some number?

Example 2. Where angles are measured in degrees, for which positive integers θ less than 90 is

$$\sin\,\theta = \sin\,\theta^2?$$

For the first example, one tool is the pigeonhole principle: *"If more than n pigeons occupy n holes, then at least one hole contains more than one pigeon,"* and its extension: *"If more than $9n$ pigeons occupy n holes, then at least one hole contains more than 9 pigeons."* Together with this principle we need to sum the integers 1 through 9, an elementary application of summing terms in an arithmetic sequence. The second example combines the trigonometric fact *"The sine of two angles will be equal if and only if either their sum is an odd multiple of 180° or their difference is an even multiple of 180°"* with facts from number theory concerning when *"The product of consecutive integers, $\theta(\theta \pm 1)$, is an even or an odd multiple of $2^2 3^2 5$."*

Problem 29 on the 44th AHSME illustrates how three very familiar facts about triangles

- *the area of a triangle is base times height,*
- *the altitude times the base yields twice the area,*
- *the triangular inequality*

and a well-known algebraic property

- *when the product of two variables is a constant, those variables are inversely proportional* can be combined to create a most interesting problem; namely, a sort of reverse triangular inequality involving altitudes

$$h_c \leq h_a + h_b$$

where h_x is the length of the altitude to the side of length x.

The following is a sample list of mathematical tools which can be used to start your mathematical workshop for competitions. Glancing over the solutions or the classification of problems in this book for recurring topics will suggest many more tools you will wish to add to your supply.

A Geometric Tool

Students might leave some geometry courses thinking the only relevant information is about congruence of triangles. This, of course, is not an accurate impression because similarity of triangles has so many applications. For example, the whole field of plane trigonometry is based on similar triangles, as is the related possibility of assigning one number, called the slope, to a line to specify its direction.

Properties of similar triangles frequent all mathematics competitions. One example of an application is known variously as the **power of the point** formula or the **property of intersecting chords** and secants:

If $\overline{AB}$ and $\overline{CD}$ are two chords of a circle that intersect (possibly after extending them) at a point P, then

$$AP \cdot BP = CP \cdot DP.$$

The proof depends on whether the intersection point P is inside or outside the circle. First, suppose P is inside the circle. Then the verticle angles $\angle APC$ and $\angle BPD$ are equal. Also, $\angle CAP = \angle PDB$ because $\angle CAB$ and $\angle CDB$ intersect the same arc of the circle. Consequently, $\triangle APC \sim \triangle DPB$, so

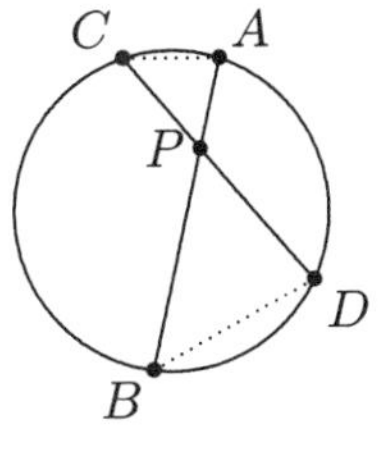

$$\frac{AP}{DP} = \frac{CP}{BP}$$

and thus

$$AP \cdot BP = CP \cdot DP.$$

C A P D B

On the other hand, suppose P is exterior to the circle. Then, because the arcs of the circle intercepted by $\angle CAB$ and $\angle CDB$ make up the whole circle, it follows that $\angle CAB$ is supplementary to $\angle PDB$. But $\angle CAB$ is also supplementary to $\angle PAC$, so $\angle PDB = \angle PAC$. Therefore, $\triangle PAC \sim \triangle PDB$. Consequently,

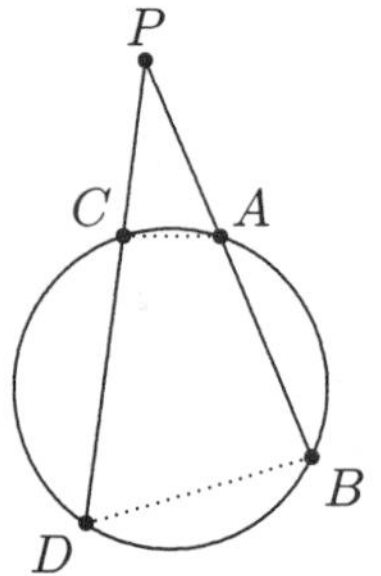

$$\frac{AP}{DP} = \frac{CP}{BP}$$

so

$$AP \cdot BP = CP \cdot DP.$$

Number Theory Tools

Something as mundane as long division provides useful results when analyzed. Whether a and b are positive integers or polynomials, the process

$$\begin{array}{r} q \\ a\overline{)\ b} \\ \vdots \\ \hline r \end{array}$$

can be expressed algebraically as

$$b = aq + r.$$

For integers, letting d and e be the greatest common factors of (a, b) and (a, r), respectively, it is very easy to show that $d \le e$ and $e \le d$, and hence $d = e$. This fact is frequently useful in competition problems. Many times we might see it expressed in the equivalent form

$$\frac{b}{a} = q + \frac{r}{a}$$

so b/a will be reduced to lowest terms if and only if r/a is similarly reduced. When this fact is put into a procedure and continued,

$$\text{gcf}(a, b) = \text{gcf}(a, r) = \cdots,$$

to obtain ever smaller pairs of integers until a remainder of 0 occurs, this procedure is known as the **Euclidean Algorithm**. [Remember that Euclid did not have electronic calculators available when he discovered and proved that there are an infinite number of primes, so he had to be clever in the tools he developed for experimentation.]

Where $p_1, p_2, p_3, \ldots, p_i$ are distinct primes, the divisors of

$$n = {p_1}^{a_1}{p_2}^{a_2}{p_3}^{a_3} \cdots {p_i}^{a_i}$$

are all numbers of the form

$$n = {p_1}^{b_1}{p_2}^{b_2}{p_3}^{b_3} \cdots {p_i}^{b_i} \quad \text{where } 0 \le b_j \le a_j \text{ for all } j = 1, 2, 3, \ldots i.$$

Since there are $a_j + 1$ choices for each b_j, **the number of divisors** of n is

$$(a_1 + 1)(a_2 + 1)(a_3 + 1) \cdots (a_i + 1).$$

The sum of the divisors of n is the product of the sum of the divisors of each ${p_i}^{a_i}$

$$(1 + p_1 + {p_1}^2 + \cdots + {p_1}^{a_1})(1 + p_2 + {p_2}^2 + \cdots + {p_2}^{a_2}) \cdots$$
$$(1 + p_i + {p_i}^2 + \cdots + {p_i}^{a_i})$$

since the terms in the expansion of this expression include each divisor d of n exactly once. Use the formula for the sum of a finite geometric sequence on each of the factors to write this formula more succinctly:

$$\left(\frac{{p_1}^{a_1+1} - 1}{p_1 - 1}\right)\left(\frac{{p_2}^{a_2+1} - 1}{p_2 - 1}\right) \cdots \left(\frac{{p_i}^{a_i+1} - 1}{p_i - 1}\right).$$

A Logarithmic Tool

Logarithms are very important in college mathematics. Calculators have replaced the use of logarithms in high school calculations. Preparation for mathematics competitions can be used to motivate the study of logarithms, whose understanding is still just as important in higher mathematics. Logarithm problems on mathematics competitions continue to encourage students to become familiar with logarithms and their properties. Besides the definition of a logarithm and the rules for the logarithms of products, quotients, and powers, another tool used in competitions is

$$\log_y x = 1/\log_x y.$$

Its proof follows directly from the definition of logarithm. Let $t = \log_y x$. Then

$$\begin{aligned} x = y^t &\iff \log_x x = \log_x y^t \\ &\iff 1 = t \log_x y \\ &\iff t = 1/\log_x y \\ &\iff \log_y x = 1/\log_x y. \end{aligned}$$

The Tool of Mathematical Induction

Problems amenable to the inductive approach might give a rule that must be applied an extremely large number of times to find the answer. **Examples:**

1. For the sequence of points in the complex plane where

$$z_1 = 0, \quad z_{n+1} = z_n^2 + i \text{ for } n \geq 1,$$

   find z_{2000}.
2. If 2000 lines are drawn in the plane with no pair parallel and no three intersecting at the same point, then into how many regions is the plane divided?
3. Find the 2000th term in the sequence

$$1, 2, 2, 3, 3, 3, 4, 4, 4, 4, 5, 5, 5, 5, 5, 6, 6, 6, 6, 6, 6, \ldots.$$

For such problems, begin as if you are going to do the problem the 'long way' (in these examples, starting at 1 and continuing step by step to 2000.) However, as you do this, keep your eyes open for patterns. This is the inductive approach as it applies to problem-solving on competitions.

When this approach is used on the first example, we find

$$z_1 = 0, \; z_2 = i, \; z_3 = i - 1, \; z_4 = -i, \; z_5 = i - 1.$$

Since each point depends only on the previous point, as soon as some point in the complex plane recurs in the sequence, we know we have entered a cyclic pattern. In this case we see that $z_{2000} = z_4 = -i$ since $z_5 = z_3$.

To apply this approach to the second problem, let $f(n)$ be the number of regions when there are n lines. A simple sketch shows that

$$f(0) = 1, \; f(1) = 2, \; f(2) = 4.$$

Don't immediately jump to the conclusion that $f(n) = 2^n$. Many sequences may start with the same three numbers, but then follow different paths.[†] When we add a third line to our sketch, we see that $f(3) = 7$, not 8. This third line cannot subdivide each of the $f(2) = 4$ regions because to do so it would have to intersect one of the first two lines twice. Add a few more lines to your sketch. Note that the nth line intersects all the previously drawn $n-1$ lines at distinct points. These intersection points divide the nth line into n pieces, each of which creates one new region by subdividing one previous region into two. Symbolically,

$$f(n) = f(n-1) + n$$

so

$$\begin{aligned} f(n) &= f(n-1) + n = [f(n-2) + (n-1)] + n = \cdots \\ &= f(1) + 2 + 3 + \cdots + n, \end{aligned}$$

from which we obtain the formula

$$f(n) = 2 + (n-1)\frac{n+2}{2} = \frac{n^2+n+2}{2},$$

so $f(2000) = 2{,}001{,}001$.

In the third example, which is related to problem 16 on the 44th AHSME, the observations that

- *the last* 2 *occurs in position number* $1+2$
- *the last* 3 *occurs in position number* $1+2+3$
- *the last* 4 *occurs in position number* $1+2+3+4$

are the inductive part of the problem. Then one must approximate a solution to

$$\frac{n(n+1)}{2} = 2000$$

which is easily accomplished by approximating the square root of $2 \cdot 2000$.

In the classroom we cover the two parts of mathematical induction, the basis of the induction and the proof of the induction step, almost as equal parts because both must be present for a valid induction argument. This argument style is also checked in the write-up of your solution on

[†] Since there are few small integers and many sequences of integers that begin with small numbers, it follows that many different sequences start with the same small numbers. This observation has been called the *Law of Small Numbers*.

free-response competitions such as the USA Mathematical Olympiad. Except for that write-up phase, the practical use of the induction method on competitions and in mathematical research has a different emphasis:

- You have worked out a number of examples for small cases. These are the cases that make you venture your guess as to the general pattern. These initial examples are usually much more than adequate as the basis for your induction.
- On a timed competition when only the answer is required, do not write out the proof of the induction step formally. Rather, when you think you have a pattern, you check to see if it persists for a few more small cases, and then try to discover a reason for the pattern.

The second example above and its solution illustrate this method beautifully.

Tools for Exponential Equations

An exponential equation is one in which the variable appears as a part of an exponent in one or more terms. If the equation to be solved can be arranged to have only one term on each side, the usual technique is to take the logarithm of both sides with respect to an appropriate base. For example, to solve

$$\frac{9^x}{9} = 3 \cdot 3^x$$

take the $\log_3$ of both sides:

$$\begin{aligned} \log_3(9^x/9) &= \log_3(3 \cdot 3^x) \\ x\log_3(9) - \log_3(9) &= \log_3(3) + x\log_3(3) \\ 2x - 2 &= 1 + x \\ x &= 3. \end{aligned}$$

Logarithms of sums or difference do not simplify. When it is impossible to avoid more than one term per side of the equation, many times a substitution will help. For example, to solve

$$\frac{9^x}{3} + 3 = 2 \cdot 3^x$$

first note that $9^x = (3^x)^2$. Therefore all variable terms can be expressed in terms of $t = 3^x$. Then

$$\frac{t^2}{3} + 3 = 2t$$
$$t^2 - 6t + 9 = 0$$
$$t = 3, \quad 3^x = 3, \quad x = 1.$$

Problem-solving is an art, not a science. That is, there is no one procedure that we can automatically use on a given type of problem. For example, after the above discussion one might blindly substitute $t = 2^x$ to solve

$$(2^x - 4)^3 + (4^x - 2)^3 = (4^x + 2^x - 6)^3$$

which is problem 20 on the 42nd AHSME. As seen in the solution, other considerations first greatly simplify this equation, and then the use of the $t = 2^x$ tool in the simplified setting easily leads to the answer.

Euler's Formula

Euler's formula for networks on the surface of simple solids is

$$r - e + v = 2$$

where r is the number of regions on the surface, v is the number of vertices, and e is the number of edges. It is very easy to verify this formula for a cube where $r = 6$, $e = 12$ and $v = 8$, for a prism with two congruent n-gons as bases where $r = n + 2$, $e = 3n$ and $v = 2n$, or for a pyramid with an n-gon as a base where $r = n + 1$, $e = 2n$ and $v = n + 1$.

The application of Euler's formula goes well beyond polyhedra. Consider the surface of a soccer ball. Each of its regions has either five sides or six, but these regions are not planar. The formula works for the corresponding polyhedron in which the faces are planar pentagons and hexagons, so the formula works for the network of regions, edges and vertices on the soccer ball.

On the soccer ball the edges are not line segments, but curves. In general, the length of the edges does not matter, and the edges need not be straight. These edges must have both ends at vertices and cannot cross other edges or vertices between their endpoints.

An application of Euler's formula with a surprising result applies to a soccer ball and its generalization. Suppose you begin with a blank sphere and draw any network on its surface subject only to these conditions:

- *All regions have a perimeter of either 5 or 6 edges.*
- *At each vertex, exactly 3 regions must come together; i.e., there are exactly 3 edges coming from each vertex.*

Note that the soccer ball fits these criteria. So does the dodecahedron because all 12 of its faces are pentagons and none or hexagons.

For this arbitrary network of five- and six-sided regions, let p be the number of five-sided retions and h be the number of six-sided regions. Then

- *The total number of regions on the sphere is* $r = h + p$.
- *The total number of sides of regions is* $6h + 5p$, *and each edge is the side of two adjacent regions, so* $e = (6h + 5p)/2$.
- *The total number of corners, summed around the perimeter of each region, is* $6h + 5p$, *and exactly three of these corners come together at each vertex of the network, so* $v = (6h + 5p)/3$.

Thus

$$r - e + v = 2$$

$$(h + p) - \left(\frac{6h + 5p}{2}\right) + \left(\frac{6h + 5p}{3}\right) = 2$$

$$p = 12.$$

That is, the only way you will ever be able to draw a network of five- and six-sided regions subject to our two criteria is if you have exactly 12 five-sixed regions. Since h dropped out of the equations, this does not imply that h is completely arbitrary. However, it is known that networks exist with 12 five-sided regions and h six-sided regions for every $h \neq 1$.

One Final Comment on Problem-Solving

This is not a psychology text and the author is not an expert in that field. On the other hand, we can all profit from reflections on some of our unsuccessful problem solving ventures. This tip is for those who have frustrated themselves working very hard for a long time on a problem, only to find that someone else solved the problem rather simply using an entirely different approach.

The longer we keep working on a problem using one method, the harder it is to think of a different method. Many very good competitors, when attempting difficult problems on mathematics competitions, will first make a brief mental or physical list of all the approaches that could be tried. Then, if what originally appeared to be the best approach does not seem to be yielding a solution, the list can be referred to for other ways to start the problem. Those who don't make this initial list of approaches before beginning the problem are those who find themselves repeatedly banging their heads against the wall at the end of the same blind alley.

In mathematics competitions as in sporting events, confidence is an extremely important factor. Those who have thoroughly studied a problem-solving book such as this will have more confidence in their next competition, as well as having learned some more mathematics.

Classification of Problems

Since many AHSME problems are interesting because they bring together diverse areas of mathematics, most problems simply do not fall into one category. The follow classification of the problems is by topics used both in the statements of the problems and in their solutions. In an effort to lead the reader to as many problems as possible on a particular topic, many problems are listed under several different topics.

The general classifications in the following are:

- Algebra, including analytic geometry, functions, and logarithms.
- Complex Numbers.
- Discrete Mathematics including "counting problems" and probability.
- Geometry.
- Number Theory including problems on bases of integers.
- Statistics.
- Trigonometry.

Within each classification sufficient subclassifications were chosen to limit the number of problems per topic.

The reference to the problems gives the number of the AHSME on which the problem occurs followed by the problem number. For example, 45-7 refers to problem number 7 of the 45th AHSME.

Algebra

Complex Numbers

Discrete Mathematics

Geometry

Number Theory

Statistics

Trigonometry